WJEC EDUQAS GCSE FOOD PREPARATION & NUTRITION

Helen Buckland

Jacqui Keepin

Consultant Editor: Judy Gardiner

HODDER
EDUCATION
AN HACHETTE UK COMPANY

ACKNOWLEDGEMENTS

Every effort has been made to trace the copyright holders of material reproduced here. The authors and publishers would like to thank the following for permission to reproduce copyright illustrations.

Section 1 © Mikael Damkier – Fotolia, Chapter 1 © Elena Schweitzer – Fotolia, Figure 1.1 © 1997 C Squared Studios/Photodisc/Getty Images/ Eat, Drink, Dine 48, Figure 1.2 © palomadelosrios - iStockphoto via Thinkstock/Getty Images, Figure 1.4 © Imagestate Media (John Foxx) / Animals, Nature & Scenics Vol 30, Figure 1.6 © Ingram Publishing Limited / Ingram Image Library 500-Food, Figure 1.7 © Stockbyte/Getty Images / Essential Produce SD153, Figure 1.8 © Jacqui Keepin, Figure 1.9 © Photodisc/Getty Images/World Commerce & Travel 5, Figure 1.10 © R A Rayworth / Alamy Stock Photo, Figure 1.11 © Getty Images/iStockphoto/Thinkstock, Figure 1.12 © yellowj – Fotolia, Figure 1.13 © winterstorm – Fotolia, Figure 1.14 © Han van Vonno – Fotolia, Figure 1.16 © Brent Hofacker – Fotolia, Figure 1.17 © vvoe – Fotolia, Figure 1.19 © danimages – Fotolia, Figure 1.21 © FOOD-pictures – Fotolia, Figure 1.22 © dream79 – Fotolia, Figure 1.23 © Stockbyte/ Photolibrary Group Ltd/ Environmental Issues DV 48, Figure 1.24 © Iain Sarjeant/ iStockphoto.com, Figure 1.26 © Igor Dutina – Fotolia, Figure 1.27 © BrandX/Getty Images / Food and Textures CD X025, Figure 1.28 © Elenathewise – Fotolia, Figure 1.29 © BrandX/Getty Images / Food and Textures CD X025, Figure 1.30 © Peter Szekely / Alamy Stock Photo, Figure 1.31 © WavebreakmediaMicro – Fotolia, Figure 1.32 © Grigoriy Lukyanov – Fotolia, Figure 1.33 © giovanni1232 - iStockphoto via Thinkstock/Getty Images, Figure 1.34 © Dave King - iStockphoto via Thinkstock/Getty Images, Section 2 © Imagestate Media (John Foxx) / Food In Focus SS67, Chapter 2 © Dea/G.Cigolini, Figure 2.1 © Photolibrary.Com, Figure 2.2 © Vikram Raghuvanshi / iStockphoto.com, Figure 2.4 © CharlieAJA - iStock - iStockphoto via Thinkstock/Getty Images/Getty Images, Figure 2.6 © kovalchuk - iStockphoto via Thinkstock/Getty Images, Figure 2.7 © Sirfuji - iStockphoto via Thinkstock/Getty Images, Figure 2.8 © Dušan Zidar – Fotolia, Figure 2.9 © Monkey Business – Fotolia, Figure 2.11 © Jules Frazier/Photodisc/Getty Images / Backgrounds Objects V08, Figure 2.13a © Thomas Francois -iStock - iStockphoto via Thinkstock/Getty Images, Figure 2.13b © Radu Sebastian - iStock - iStockphoto via Thinkstock/Getty Images \, Figure 2.13c © ibaki - iStock - iStockphoto via Thinkstock/Getty Images, Figure 2.13d © Zoonar/P.Malyshev - iStock - iStockphoto via Thinkstock/Getty Images, Figure 2.14 © Gail Philpott / Alamy, Figure 2.15 © Biophoto Associates/Science Photo Library, Figure 2.16 © 1997 Siede Preis Photography/Photodisc/Getty Images/ Eat, Drink, Dine 48, Figure 2.17 © winston – Fotolia, Figure 2.18 © matka_Wariatka – Fotolia, Figure 2.19 © Fuse - iStockphoto via Thinkstock/Getty Images, Section 3 © BrandX/Getty Images / Food and Textures CD X025, Chapter 3 © Kochergin – Fotolia, Figure 3.2 © Okea – Fotolia, Figure 3.3 © 1997 John A. Rizzo/Photodisc/Getty Images/ Eat, Drink, Dine 48, Figure 3.4 © BrandX/Getty Images / Food and Textures CD X025, Figure 3.5 © Maridav – Fotolia, Chapter 4 © Maksim Shebeko – Fotolia, Figure 4.1 © Crown Copyright (covered by the Open Government Licence), Figure 4.2 © ranplett/ iStockphoto.com, Figure 4.3 © alinamd – Fotolia, Figure 4.4 © pilipphoto – Fotolia, Figure 4.5 © Andrey Starostin – Fotolia, Figure 4.6 © Floortje/iStockphoto.com, Figure 4.7 © Frédéric Prochasson – Fotolia, Figure 4.8 © mariemilyphotos – Fotolia, Figure 4.9 © . shock – Fotolia, Figure 4.11 © Monkey Business – Fotolia, Figure 4.13 © volff - Fotolia, Figure 4.15 © digitalskillet/ iStockphoto.com, Figure 4.16 © S.HarryPhotography - Fotolia.com, Figure 4.17 © Michael Neelon(misc) / Alamy Stock Photo, Figure 4.18 © RusN - iStockphoto via Thinkstock/Getty Images, Figure 4.19 © The Photo Works / Alamy Stock Photo, Figure 4.20 © Jules Frazier/Photodisc/Getty Images / Backgrounds Objects V08, Figure 4.21 © tomalu – Fotolia, Chapter 5 © Brian Jackson – Fotolia, Figure 5.1 © svl861 – Fotolia, Figure 5.3 © Comstock - iStockphoto via Thinkstock/Getty Images, Section 4 © DEX Images Images/ Photolibrary Group Ltd/Getty Images / Food Index 02 Moment FDS16502, Chapter 6 © FOOD-pictures – Fotolia, Figure 6.1 © Sergej Razvodovskij – Fotolia, Figure 6.2a © KucherAV - iStockphoto via Thinkstock/Getty Images, Figure 6.2b © Merrimon - iStockphoto via Thinkstock/Getty Images, Figure 6.3 © Diana Miller/Cultura/Science Photo Library, Figure 6.4 © Alexey Stiop – Fotolia, Figure 6.5 © ALLEKO - iStockphoto via Thinkstock/Getty Images, Figure 6.6 © Owen Price/ istockphoto, Figure 6.7 © ALLEKO - iStockphoto via Thinkstock/Getty Images, Figure 6.8 © Nancy R. Cohen/Photodisc/Getty Images/ Eat, Drink, Dine 48, Figure 6.9a © Soniacri - iStockphoto via Thinkstock/Getty Images, Figure 6.9b © The_Pixeltree - iStockphoto via Thinkstock/ Getty Images, Figure 6.12 © Upyanose/ Thinkstock/iStockphoto/Getty Images, Figure 6.14 © Pål Espen Olsen/ iStockphoto.com, Figure 6.15 © kitti bowonphatnon – Fotolia, Figure 6.16 © Tuned_In - iStockphoto via Thinkstock/Getty Images, Figure 6.17 © Melba Photo Agency / Alamy / Ingredients CD0163D, Figure 6.18 © Awe Inspiring Images – Fotolia, Figure 6.19 © contrastwerkstatt – Fotolia, Figure 6.20a © Paul_Brighton - iStockphoto via Thinkstock/Getty Images, Figure 6.20b © Tatiana Volgutova - iStockphoto via Thinkstock/Getty Images, Figure 6.20c © Picsfive -iStockphoto - iStockphoto via Thinkstock/Getty Images, Figure 6.21 © SvetaVo - iStockphoto via Thinkstock/Getty Images, Figure 6.22 © Elena Elisseeva - iStockphoto via Thinkstock/Getty Images, Figure 6.23 © Kevin Wheal / Alamy Stock Photo, Figure 6.24 © marrakeshh - iStockphoto via Thinkstock/Getty Images, Figure 6.25 © Sasajo – Fotolia, Figure 6.26 © olgakr - iStockphoto via Thinkstock/Getty Images, Figure 6.27a © Picsfive - iStockphoto via Thinkstock/Gety Images, Figure 6.27b © Art Directors & TRIP / Alamy Stock Photo, Figure 6.27c © Ian Francis / Alamy Stock Photo, Figure 6.28 © LOU63 - iStockphoto via Thinkstock/Getty Images, Figure 6.29 © SMcTeir/Hodder Education, Figure 6.30 © jean-marie guyon -iStock via Thinkstock/Getty Images, Figure 6.31 © PhotoAlto/Laurence Mouton/Getty Images, Chapter 7 © poplasen - Fotolia.com, Figure 7.2 © Mediscan / Alamy Stock Photo, Figure 7.3 © marilyn barbone – Fotolia, Figure 7.4 © dpullman - iStockphoto via Thinkstock/Getty Images, Figure 7.6 DAJ - iStockphoto via Thinkstock/Getty Images, Figure 7.8 © Cultura RM / Alamy Stock Photo, Figure 7.9 © gabuchia - iStockphoto via Thinkstock/Getty Images, Figure 7.10 © Steven Depolo/ Flickr, Figure 7.11 © popovaphoto - iStockphoto via Thinkstock/Getty Images, Figure 7.12 © Kondor83 - iStockphoto via Thinkstock/Getty Images, Figure 7.13 © Cultura RM / Alamy Stock Photo, Figure 7.14 © Flonline digitale Bildagentur GmbH / Alamy Stock Photo, Figure 7.15 © Claus Mikosch – Fotolia, Section 5 © stocksolutions – Fotolia, Chapter 8 © blas – Fotolia, Figure 8.1 © Imagestate Media (John Foxx) / Agriculture SS90,

ACKNOWLEDGEMENTS

INTRODUCTION TO WJEC EDUQAS (9–1) GCSE IN FOOD PREPARATION AND NUTRITION

This book has been written to help you master the skills, knowledge and understanding you need for the WJEC Eduqas GCSE (9–1) GCSE in Food Preparation and Nutrition.

The course will provide you with the knowledge, understanding and skills you need to prepare and cook food safely using a wide range of different food commodities, cooking skills and techniques and equipment. It will help you to understand:

- the relationship between diet and health and the effects of a poor diet on health
- how a range of different factors affect our food choices (including economic, environmental, ethical and cultural influences, as well as food availability and food production processes)
- the functional and nutritional properties, working characteristics and sensory qualities of foods and how food science principles can be applied
- how to prepare, process, store and cook food safely
- different culinary traditions and the ingredients and cooking techniques they use.

The course is divided into six areas of content:

1 Food commodities
2 Principles of nutrition
3 Diet and good health
4 The science of food
5 Where food comes from
6 Cooking and food preparation.

Each of these areas will be assessed in the Principles of Food Preparation and Nutrition written examination.

You will also complete two non-examination assessments: Food Preparation and Nutrition in Action, which will by marked by your teacher and moderated by WJEC Eduqas.

Component	Assessment type	Format	Time	% of qualification
Principles of Food Preparation and Nutrition	Written exam (paper based or online version also available)	Two sections of compulsory questions to assess the six areas of content listed above: • Section A: questions based on stimulus material • Section B: structured, short and extended response questions	1 hour 45 minutes	50%
Food Preparation and Nutrition in Action	Non-examination assessment: Assessment 1	**The Food Investigation Assessment** A scientific food investigation which will assess knowledge, skills and understanding relating to the scientific principles underlying the preparation and cooking of food.	8 hours	15%
	Non-examination assessment: Assessment 2	**The Food Preparation Assessment** Preparing, cooking and presenting a menu to assess knowledge, skills and understanding relating to the planning, preparation, cooking and presentation of food.	12 hours	35%

Summary of assessment

More information to help you prepare for these assessments can be found in Chapters 13–15 of this book.

HOW TO USE THIS BOOK

The book is divided into eight sections:

- **Sections 1–6** cover the six areas of subject content you will be tested on in the Food Preparation and Nutrition written exam.
- **Section 7 Food preparation and nutrition in action** includes chapters on the two non-examination assessments: The Food Investigation Assessment and The Food Preparation Assessment. It explains the format each assessment will take, how many marks are available for the different aspects of each assessment, and includes example work for assessment tasks similar to those you will need to complete.
- **Section 8 Principles of food preparation and nutrition** is designed to help you to prepare for the Food Preparation and Nutrition written exam. It includes details on the format of the exam paper and types of questions you may encounter. It also includes practice questions, with sample answers and mark schemes.

Throughout the book, you will find the following features:

Learning objectives

Learning objectives are provided at the start of every chapter and tell you what you should know and understand by the end of the chapter.

Activity

Activity boxes include short activities your teacher may ask you to complete to help you develop your knowledge and understanding of a topic.

Practical activity

Practical activities are food preparation and cooking tasks that will provide you with opportunities to demonstrate the different food preparation skills and techniques, as well as allowing you to apply your subject knowledge and understanding in a practical context and see how it is relevant to practical food preparation and cooking.

KEY WORDS

Key words are provided at the end of each chapter or topic and define all of the important terms you will need to know and understand.

Stretch & challenge

Stretch and challenge activities are extension tasks that will help you to develop your knowledge and understanding of a topic further. They may ask you to complete further research into a topic, or consider some of the more challenging aspects of the course.

KEY POINTS

- **Key points** are provided at the end of each topic and summarise all of the important knowledge you will need about that topic.

Exam practice questions

Exam practice questions are given at the end of every topic or chapter. These will help you to test your knowledge and understanding in preparation for the written exam.

SECTION 1

Food commodities

This section contains the following chapters:

Chapter 1 Food commodities

CHAPTER 1
Food commodities

Learning objectives

In this chapter you will learn about:
- the value of different commodities in the diet
- the features and characteristics of each commodity, including how to store them correctly to avoid food contamination
- the working characteristics of each commodity
- the origins of each commodity.

For each food commodity you need to be able to:
- experiment with the commodity to explore chemical and physical changes as a result of given actions
- consider complementary actions of each commodity within a recipe
- cook dishes using the commodities.

This chapter contains information about food commodities that should be used and applied to the other chapters in the book. The commodities that are discussed in this chapter will be referenced throughout the other chapters. It is expected that you will use these food commodities during practical activities throughout the course and in your Non-Examination Assessment (NEA).

Bread, cereals, flour, oats, rice, potatoes and pasta

Bread

Bread is a **staple** food. This means it is eaten regularly and makes a significant contribution to the diet.

Bread is a very versatile food. It can be served and eaten on its own as well as being used to produce a wide range of food products that can form part of a main meal, or lighter types of meals such as breakfast and lunch, as well as snacks. It can be sweet or savoury.

There are many different types and styles of bread available in shops, bakeries and cafés due to the multicultural society we live in, which allows us to experience exciting food opportunities. The varieties of breads available have different textures; for example, they can be soft, chewy, crusty and airy.

To make bread more interesting and varied, different ingredients are often added to add flavour (e.g. cheese and herbs), as well as different fillings and toppings (e.g. poppy seeds).

To make bread, the main ingredients used include:

- Flour
- Salt
- Yeast
- Liquid (this is normally water, but milk can be used to give the dough richness).

A small amount of sugar and a fat can also be added.

The ingredients are combined to form a dough, which is normally kneaded, proved, shaped and baked.

You will learn more about shaping and finishing dough in Chapter 11 Preparation and cooking techniques.

Bread can be **leavened** (risen using a raising agent such as yeast) or **unleavened** (flat).

You will learn more about using raising agents in Chapter 11 Preparation and cooking techniques.

Figure 1.1 A range of different bread products

Nutritional value

Bread is a regarded as a nutritious food. It is a good source of carbohydrates, protein, B-group vitamins and the minerals calcium and iron. If wholemeal flour is used in the making of the bread it will also be a good source of dietary fibre (NSP).

Bread fits into the 'bread, rice, potatoes, pasta and other starchy carbohydrates' section of the Eatwell Guide. This is the section coloured yellow. It is recommended that one third of the diet should be made up of foods from this section.

You will learn more about the Eatwell Guide and recommended guidelines for a healthy diet in Chapter 4 Plan balanced diets.

Storage of bread

How bread should be stored depends on the type of bread made or bought. Fresh bread can become stale quite quickly. Some manufacturers will add preservatives during the making process to give bread a longer shelf life.

- Fresh bread from a baker is best stored in a bread box or alternatively stored in a paper bag with the top closed. This prevents loss of moisture when exposed to air.
- Supermarket-style loaves are stored in plastic, which will keep them fresh for a period of time. If there is moisture present it can make the bread go mouldy and become unsafe and unpleasant to eat. The bread should be stored in a cool, dry environment.
- Freezing bread is a suitable method of storage. Slices can be wrapped individually and wrapped in plastic to keep the bread fresh. It can be stored for two months or longer and taken out and defrosted when required.
- Bread should not be stored in the refrigerator as this can make the bread become very dry and change its flavour.

You will learn more about storing foods correctly in Chapter 7 Food spoilage.

Cereals

Figure 1.2 Different breakfast cereals

To many people, the term 'cereal' means breakfast cereals. These are often eaten for a nutritious start to the day by people of all ages. Breakfast cereals are made from cereal grains that could have been puffed, shredded or flaked to achieve the desired cereal product. For example, Rice Krispies are toasted, crisped grains of rice. Cereals are eaten every day in many different forms and in a wide variety of products.

Cereals are grown all over the world. In the UK, some of the cereals we grow include:
● Wheat ● Barley ● Oats ● Rye.

For cereals to grow in the UK they have to be well-suited to the temperate climate.

Cereals are actually **cultivated** grasses, which are grown for their nutritious edible seeds, known as cereal grains. Most cereal grains tend to be the same in their structure but will have different textures, colours, shapes and sizes.

You will learn more about how cereals (crops) are grown in Chapter 8 Food provenance.

Activity ✏

1 The picture below shows a variety of cereal plants. Name each of the cereal plants shown.

2 Name a food product that is made using each of the cereal plants shown.

Figure 1.3 A range of different cereal plants

Nutritional value

Cereals provide a valuable source of energy in the diet, as well as other nutrients if the **wholegrain** is used. These include:

- fibre
- carbohydrates
- protein (LBV)
- B vitamins
- vitamin E
- fat
- iron

Many of the cereals grown are **processed** before they are used or eaten. For example, when wheat is grown in the field it is a raw food material; this is known as a primary source. It is then processed into an ingredient that can be used to produce other food products, for example flour.

You will learn more about how cereals are processed in Chapter 9 Food manufacturing.

Storage of cereals and cereal products

Cereals can become stale, **perish**, lose flavour, develop odours and become contaminated with bacteria and mould, which could lead to food poisoning if not stored correctly. To store cereals safely:

- Keep them away from moist environments, preferably store in a cool, dry place.
- Store them in an airtight container to maintain freshness, or fold over and seal the packaging.
- Always check the use-by date before use.
- Keep new and older cereals separate; do not mix. Always use the older cereal first and do not 'top up' with new cereals.

Flour

Flour comes from different types of cereals, for example rye and wheat.

Wheat flour is one of the main flours produced. There are different strengths of wheat flour depending on its uses:

- Strong flour is used in bread making and comes from winter wheat, which is a hard wheat.
- Weak flour is used in cake and biscuit making and comes from spring wheat.

KEY WORDS 🔑

Cultivated: raised or grown on a farm or grown under controlled conditions.

Perish: to spoil or to go bad.

Processed: changes made to a raw material to make it more edible.

Wholegrain: the whole seed in its natural state (none of the layers have been removed).

KEY POINTS: CEREALS 🎯

- Cereals are cultivated grasses.
- Wheat is a common cereal grown and used in the UK.
- Cereals provide a valuable source of energy in the diet.
- Many cereals have to be processed before they can be used or eaten.
- Cereals are best stored in an airtight container in a cool, dry place.

Nutritional value

Strong flours have a high protein content; this becomes gluten when the flour is mixed with a liquid during the making of a product. Gluten will give a dough elasticity and stretch. When baked this will set and form a solid structure for a product.

You will learn more about gluten formation in Chapter 6 The effect of cooking on food.

There are many different varieties of flour available for people to buy. Flour is the main ingredient in a variety of baked products as it fulfils many different functions.

In the UK, white flour has to be **fortified** with the minerals calcium and iron and the B vitamins thiamine and niacin; this is because they have been lost during processing.

You will learn more about how flour is processed in Chapter 9 Food manufacturing.

The table below shows **the extraction rate** of different flours. The extraction rate identifies how much of the wholegrain has been used to make the flour.

Type of flour	Extraction rate	What has been removed
White flour	extraction rate 70–75%	The bran, germ, fat and some of the minerals
Wholemeal flour	extraction rate 100%	No removal, the wholegrain is used

Table 1.1 Types of flour

Storage of flour

To store flour safely and to maintain its quality:

- Store in its original sealed packaging in a cool, dry cupboard.
- Store in an airtight container to prevent **weevils**.
- Never mix old flour in a bag with new flour.
- Check use-by dates before use. Wholemeal flour does not keep as well as white flour, as the oils from the germ and bran can become **rancid** with age.

Activity

1 Carry out some research on the different flours listed below. Find out what the differences between them are. Present your findings in a chart like the one shown below.

Type of flour	Main differences
Gluten-free flour	
Self-raising flour	
Rye flour	

Table 1.2 Types of flour

2 Name two different products each type of flour could be used to make.

3 In the UK, white flour has to be fortified. Explain what the term 'fortified' means.

Oats

Oat grains have a protective husk covering them; this has to be removed before oats can be used as a food.

- Oats can be ground into a fine flour, which can then be used to make products such as muffins, biscuits and scones.
- Oat flour can be made by processing rolled oats in a food processor until the consistency is fine and powdery and then stored in an airtight container.
- Oats can be rolled or crushed – this will produce oatmeal.
- Porridge is made using rolled oats. This type of oats can be used to make flapjacks.
- Jumbo oat flakes can be used in the making of baked goods, or in breakfast cereals such as granola and muesli. They can give a product extra texture due to the bigger flake.

Nutritional value

Oats are a very nutritious cereal. The main nutrient provided is carbohydrates, but they also contain some protein, fat, calcium, and iron and B group vitamins.

Wholegrain oats can provide slow-release energy; this means the energy release will be over a longer period of time when compared to sugary foods, which will give a quick energy release over a shorter period of time. Including wholegrain oats in a slimming or diabetic diet is very beneficial due to the high carbohydrate content which gives a prolonged feeling of fullness.

Storage of oats

- Store in a cool, dry area.
- Once opened, store in an airtight container.

Figure 1.4 Oat cereal grain

Rice

Rice is one of the most popular staple foods eaten by the world's population. It is a very versatile commodity because it can be used to make both sweet and savoury dishes.

Rice is served as part of a meal to provide bulk and a feeling of fullness. It is quick to cook and is a good store cupboard ingredient as it has a long shelf life and is easy to store.

Rice can be quite bland in flavour. This can be improved by cooking it with flavoursome ingredients such as garlic and herbs, or by cooking the rice in stock instead of water. It can also provide a balanced taste to a meal if it is being served with strong or spicy flavoured dishes such as a curry or chilli-style dishes.

You can cook rice using different cooking methods:
- boiling on the hob
- baking in the oven
- stir-frying once cooked.

When rice is **harvested** the grains are covered in a thick outer husk. This is removed during processing.

You will learn more about processing in Chapter 9 Food manufacturing.

There are many different varieties of rice available in supermarkets and it is sold in a variety of different forms, for example boil-in-the-bag, easy cook and pre-cooked.

Rice can be short grain or long grain and most types are available as brown or white rice. Some of the different varieties of rice and their uses are shown in the table below.

Varieties of Rice	Uses
Short grain rice	
Arborio	Risotto
Pudding	Sweet rice pudding
Long grain rice	
Basmati	Served with many Indian-style dishes
Carolina	Savoury rice dishes

Table 1.3 Varieties of rice and their uses

Nutritional value

Rice is regarded as the poorest of all cereal foods in relation to its protein, fat and mineral content, but it is an excellent source of energy.

Storage of rice

To store uncooked rice:

- Store in a cool, dry area.
- Once opened, store in an airtight container.

It is recommended that cooked rice should not be stored and reheated, as this can lead to food poisoning. Once cooked, rice becomes a **high risk food**. If it is necessary to store cooked rice:

- Store above 65°C for no longer than two hours.
- Rinse in cold water immediately after cooking, chill and refrigerate.

You will learn more about food poisoning in Chapter 7 Food spoilage.

Wild rice

Brown rice

Basmati rice

Easy-cook rice

Figure 1.5 Four different types of rice

Potatoes

There are many different varieties of potatoes grown in the UK. Some examples of these include Maris Piper, King Edward and Desiree. Sweet potatoes are also a popular choice as an alternative to traditional potatoes.

The part of the potato plant we eat is called the **tuber**. Potato tubers can come in a variety of colours; we are most familiar with red and white potatoes. When we make a choice between a red or white potato it is often related to the taste and the type of recipe being prepared.

Figure 1.6 Red and white potatoes

The variety of the potato used when preparing meals and dishes can result in very different textures and outcomes. Cooked potatoes can be floury, sticky or waxy and granular. This is due to the potato cell changing during the cooking process.

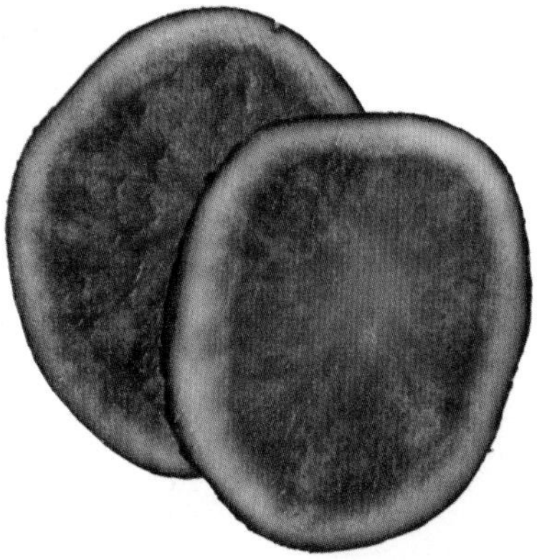

All the different varieties of potatoes have the same structure:

- Outer layer is the **skin.**
- The **flesh** is the area under the skin.
- The **pith** is the watery core, the innermost part.

Potatoes are regarded as a traditional staple food. In the UK, they are often eaten as the main accompaniment to dishes. They can be prepared and cooked in a variety of ways:

- baked   - roasted   - fried   - boiled.

Figure 1.7 The structure of a potato

You will learn more about these methods in Chapter 11 Preparation and cooking techniques.

Nutritional value

Potatoes can be a good source of vitamin C due to the amount eaten and the number of times they are eaten. They provide carbohydrates in the form of starch, some calcium and a small amount of B group vitamins. They also contain water.

Storage of potatoes

- Potatoes can be stored in hessian bags, paper bags or in racks. They should be stored in a cool, dry, dark, airy place.
- Storing potatoes in a light environment can cause them to turn green. This should be removed before cooking as the green part is **toxic.**
- Potatoes should not be stored in plastic bags as this can cause them to sweat and rot.
- Storing potatoes in a refrigerator can affect the taste and cause discolouration when they are cooked.

Activity

Potatoes are used in the making of a wide range of savoury products. Produce a mind map to show how many different savoury products they can be used in.

Practical activity

1 Potatoes can be cooked by different methods. Choose three different cooking methods and prepare potatoes using these cooking methods.
2 Ask people in your class to taste test them and identify which method is best and why. You could record your findings using sensory profiles (refer to Chapter 10 Factors affecting food choice, page 184, to find out more about sensory profiles).
3 Analyse the nutritional value of each method used, to find out which would provide the highest value.

KEY WORDS

Tuber: a thickened underground part of a stem or root from which new plants can grow.

Skin: the outside layer or covering.

Flesh: the area under the potato skin.

Pith: the watery core.

Toxic: containing a harmful substance.

KEY POINTS: POTATOES

- Potatoes are a staple food in the UK.
- There are many different varieties of potatoes.
- The tuber is the part of the potato plant which we eat.
- All potatoes regardless of variety have the same structure.
- Potatoes can be a good source of vitamin C.
- Potatoes should be stored in a cool, dry, dark, airy place.

Pasta

Pasta is made from strong wheat known as **durum wheat**. This type of wheat contains more protein than common wheat. During the milling process the wheat produces semolina; this is the coarsest grade of the starchy **endosperm**.

You will learn more about how wheat is grown and processed in Chapter 8 Food provenance and Chapter 9 Food manufacturing.

To make pasta, water is added to form a dough, which can be shaped or **extruded** (forced through an opening in a shaped plate and then cut to a specific size) to produce the type of pasta required.

Other ingredients that can be added during the making of the pasta dough include eggs, oil, salt and various flavourings.

Different shapes, sizes and styles of pasta are widely available to buy in shops. Various colours of pasta are also sold:

- Green pasta is made using spinach, which provides the colour as well as some flavour.
- Red pasta is made using tomato paste.
- Squid ink pasta or black pasta is dark grey, almost black in colour and is made using, as the name suggests, squid ink. This can sometimes give the pasta a mild seafood flavour.

Figure 1.8 Dried and cooked pasta

Dried pasta is popular due to its long shelf life and versatility; it can be combined with many other ingredients. When dried pasta is cooked it changes to a lighter colour and increases in size as it absorbs the cooking liquid. The photograph shows the differences between dried and cooked pasta.

You will learn more about preparing and cooking pasta in Chapter 11 Preparation and cooking techniques.

Storage of pasta

- Dried, uncooked pasta can be stored in its original packaging. Once opened, store in an airtight container in a cool, dry place away from strong odours.
- Fresh pasta must be stored in a refrigerator.
- Homemade pasta must be allowed to dry and then stored in an airtight container in the refrigerator.
- Fresh and homemade pasta can be frozen.
- Cooked pasta should be stored in an airtight container in the refrigerator. Rinsing with cold water after cooking will stop it sticking together.

Activity

1 Produce a mind map to show the many different types of pasta that are available.
2 Make a list of six different sauces that could be served with pasta.

Practical activity

Find a recipe and make your own pasta dough. Make two different types of pasta with the dough: one shaped pasta and one filled pasta.

KEY WORDS

Durum wheat: wheat which contains a high level of protein.

Endosperm: a tissue produced inside seeds of flowering plants around the time of fertilization. It surrounds the embryo and provides nutrition in the form of starch, though it can also contain oils and protein.

Extruded: a soft mixture being squeezed under pressure through a shaped die (tool, part of a machine) into strips and cut into even pieces.

KEY POINTS: PASTA

- Pasta is made from durum wheat and water.
- Many different shapes, styles and sizes of pasta are sold.
- Pasta can be made in different colours by adding extra ingredients.
- Pasta can be bought fresh or dried.
- Dried pasta increases in size when cooked.

Exam practice questions

1 List **three** of the main nutrients found in cereals. [3 marks]
2 Explain why we should choose cereal products that are of a wholegrain variety when making dishes. [3 marks]
3 State the type of flour used when making bread. Explain why it is important to use this type of flour. [4 marks]
4 Explain why a person who is following a slimming diet should consider having an oat-based breakfast cereal to start the day. [3 marks]
5 Rice can be cooked by baking it in the oven. Give **two** detailed advantages of using this method to cook rice. [4 marks]
6 Explain why it is important to store potatoes correctly. [4 marks]
7 Explain why dried pasta is considered to be a good store cupboard ingredient for a large family. [5 marks]

Fruit and vegetables

Fruit

KEY WORDS 🗝

Seasonal: the times of year when a food is at its peak, in terms of harvest or its flavour and sometimes cost.

There are many different types of fruits available in supermarkets that are locally, regionally, nationally or internationally grown. Many fruits are **seasonal**. This means they are not available all year round. Many fruits are imported from different countries so that they are available all year round due to customer demand.

The Eatwell Guide recommends that one third of our diet should be made up of fruits and vegetables and we should try to eat a wide variety of them.

You will learn more about the Eatwell Guide and recommended guidelines for a healthy diet in Chapter 4 Plan balanced diets.

Fruits and vegetables are very nutritious. It is recommended that at least five portions of fruits and vegetables are eaten every day and a variety of colours and types are eaten to help meet different vitamins and minerals requirements. This group of foods provides:

- carbohydrates
- dietary fibre
- vitamins A, C and E
- various minerals.

They are regarded as a low fat food.

There are four main groups of fruits. The table shows the different groups of fruits, examples of fruits that fit the group and how to store them.

Groups of fruits	Example fruits	Storage
Citrus	Oranges, lemons, grapefruits, limes	Cool dry place. Some citrus fruits can be refrigerated.
Hard fruits	Apples, pears	Keep out of direct sunlight, at room temperature. Can be refrigerated.
Soft or berry fruits	Strawberries, raspberries, blackberries	Keep refrigerated. Remove to serve at room temperature
Stone fruits	Plums, cherries, nectarines, peaches	Keep refrigerated. Store in fruit bowl at room temperature for faster ripening.

Table 1.4 Different groups of fruits

There are some fruits that do not fit into any of groups identified in the table. These fruits include bananas, pineapple, mango and passion fruit. Sometimes supermarkets will class these types of fruits as tropical or exotic.

Fruits are available in many different forms, for example canned, frozen and dried. Before a fruit can be eaten raw it needs to ripen; this process makes fruit attractive to eat. Many changes take place when fruit ripens; some of these include colour change, texture change and a taste change when the sweetness develops.

You will learn more about how foods are grown and processed in Chapter 8 Food provenance and Chapter 9 Food manufacturing.

Figure 1.9 Examples of some tropical fruits

Vegetables

Vegetables are grouped according to the different parts of the plant they represent. These can be grown above or below the ground. The table shows the different groupings of vegetables.

Vegetable group	Examples of vegetables	Above or below ground
Roots	Beetroot, carrots, swede	Below
Bulbs	Onions, leeks, spring onions	Below
Tubers	Potatoes, sweet potatoes, yams	Below
Stems	Asparagus, celery	Above
Leaves	Different cabbage varieties, Brussels sprouts, iceberg lettuce	Above
Flowers	Cauliflower, broccoli	Above
Fruits and seeds	Peas, courgettes, aubergine	Above
Fungi	Mushrooms	Above

Table 1.5 Different groupings of vegetables

The structure of a vegetable is a collection of cells which are made of **cellulose**. The type of vegetable and the age of the vegetable can mean the structure varies.

Vegetable cells contain a high amount of water; this helps to keep them firm. If they start to lose water the cells will start to lose their firmness and become limp and flabby.

Just like fruits, vegetables are available in many different forms: dried, canned and frozen.

Nutritional value

We are encouraged to eat a wide variety of different vegetables as part of our daily diet. They can be eaten as part of a main meal or as a snack throughout the day. Many vegetables can be eaten raw; this increases their nutritional value as cooking them can destroy or reduce some of the nutrients found in them.

Storage of vegetables

- Salad and some green vegetables can be stored in the refrigerator to keep them fresh.
- Most other vegetable types should be stored in cool, dry, well-ventilated areas.
- Most vegetables are best eaten as soon as purchased to avoid nutrient and flavour losses.

KEY POINTS: FRUITS AND VEGETABLES

- Fruits and vegetables are edible plants or parts of plants.
- Fruits and vegetables should make up one third of our daily diet.
- At least five portions should be eaten a day to help achieve a healthier diet.
- Many vegetables can be eaten raw and so are ready to eat and require little or no preparation.
- They are available in many different forms.
- Fruits and vegetables are very nutritious, providing good sources of vitamins A, C and E.
- Cooking vegetables can reduce their nutritional value.

KEY WORDS

Cellulose: insoluble substance which makes up most of plants like vegetables, cell walls.

Practical activity

1 Find a recipe for a vegetable soup and make the soup showing the following preparation skills: peeling, grating, slicing and dicing. Use the **bridge hold** and **claw grip** (see Chapter 11, page 214) when executing the knife skills.
2 When the soup is cooked, divide it into two separate saucepans. Keep one half as a chunky soup and blend the other half using a blender to make a smooth soup.
3 Record the differences in taste and texture.

Exam practice questions

1 Name **three** different forms of peas that could be bought in a supermarket. [3 marks]
2 Suggest **three** different meals that could be made using potatoes. [3 marks]
3 It is recommended that we eat at least five portions of fruit and vegetables a day. Give **two** detailed reasons for following this recommendation. [4 marks]
4 Explain the benefits of using strawberries when they are in season to make a dessert. [4 marks]

Milk, cheese and yoghurt

Milk

In the UK we mostly consume cow's milk. Other milks are available as an alternative, for example goat's milk and soya milk.

Milk has to be treated to make it safe to drink. This is normally through heat treatment. Harmful **bacteria** are destroyed during the treatment. This also will give milk a longer shelf life.

There are many different ways to heat treat milk. Two examples are listed below.

- **Pasteurisation:** HTST is the normal method. This means 'High Temperature Short Time'. Milk is heated for 15 seconds to a temperature of 72°C. It is then cooled rapidly to below 10°C and bottled or put into cartons. It must be stored in a refrigerator.
- **Ultra Heat Treated (UHT):** Milk is heated rapidly for 1 second to 132°C and then rapidly cooled and packed under sterile conditions. This milk has a long life and can be stored unrefrigerated for about six months. Once opened it must be treated the same as fresh milk.

Figure 1.10 UHT Milk

Types of milk

There are many different types of milk sold in shops. The most popular ones include:

- **Whole milk:** this is full-fat milk with 3.9 per cent fat content. It has a blue cap. This milk is recommended for young children, as they are still growing.
- **Semi-skimmed milk:** this is half-fat milk with 1.7 per cent fat content. It has a green cap.
- **Skimmed milk:** this is low-fat milk with 0.1–0.3 per cent fat content. It has a red cap.
- **Evaporated milk:** this is **concentrated**, sterilized canned milk that has been heated to reduce the liquid content and kill bacteria.
- **Condensed milk:** this is evaporated milk which has sugar added to it. The sugar helps to preserve the milk.
- **Dried milk powder:** this powder is made by taking away the water from the milk and applying heat to dry it. It is mixed with water when used and must be stored as fresh milk.
- **Alternative milks:** These are milks that are dairy-free. Examples include soya, oat and rice milk.

Figure 1.11 Different types of milk

Nutritional value

Milk is referred to as a **'complete food'** as it provides many of the nutrients that are needed for health. It is designed to be the only food that a baby mammal needs for the first weeks of its life.

Milk	
Nutrient provided	**Nutrient information**
Protein	High biological value
Fat	Mainly saturated form
Carbohydrate	In the form of lactose
Minerals	Calcium, phosphorus, potassium and minimal iron
Vitamins	A and D, B vitamins, little vitamin C
Water	High-volume content

Table 1.6 Nutritional value of milk

Storage of milk

- Fresh milk is perishable and should be consumed within a few days of opening and before the use-by date.
- Milk should be refrigerated between 0 and 5°C.
- UHT milk can be stored in a cool, dry place. Once opened, store in the refrigerator and treat as fresh milk.
- Keep milk containers closed and store away from strong-smelling food items in the fridge, as milk can pick up these odours.

KEY WORDS

Pasteurisation: a method of heat treating milk using a high heat and only seconds in time.

Ultra Heat Treated (UHT): a method of heat treating milk to kill bacteria and extend the shelf life.

Bacteria: micro-organisms that contaminate food and make it unfit to consume.

Concentrated: a richer, more intense food source due to removal of water.

Lactose: A sugar from carbohydrates found in milk.

KEY POINTS: MILK

- Milk is heat treated to make it safe to drink and use.
- Pasteurisation and UHT are two methods used to heat treat milk.
- Whole milk, semi-skimmed and skimmed milk are types of milk sold in shops.
- Milk is regarded as a complete food.

Activity

1 Explain the differences between whole milk and skimmed milk.
2 State the advantages of buying UHT milk instead of fresh milk.
3 Name two sweet and two savoury dishes that are made using milk.

Cheese

Cheese could be described as a solid or semi-solid form of milk. It is also sometimes referred to as a **fermented** dairy food.

There are many different varieties available both in the UK and internationally. The table below shows some of the cheeses that are made by different methods.

Type of cheese	Examples
Hard pressed cheese	Cheddar, Gruyère, Leicester
Soft cheese (sometimes called ripened)	Camembert, Brie, Goat's
Unripened cheese (sometimes called a soft, fresh cheese)	Cottage cheese, Cream cheese, Mascarpone, Mozzarella, Feta
Blue veined cheese	Stilton, Gorgonzola, Danish blue
Processed cheese	Cheese slices and cheese spreads

Table 1.7 Types of cheese

Nutritional value

As cheese is made from milk solids its nutritional value is very similar to milk. However, it is a much more concentrated food as most of the water content is lost during manufacture. The amount of the nutrient will depend on the type of cheese. Cheese will provide the following nutrients:

- High biological value (HBV) protein
- Calcium – cheese is a good source of calcium
- Phosphorous and sodium – these amounts can vary depending on a number of factors
- It is a good source of vitamin A and provides some vitamin D and vitamin B – these amounts can also vary
- A high fat content – the amount of fat will depend on the type of milk that was used to make the cheese and the type of cheese.

Figure 1.12 **Different types of cheese**

You will learn more about the processing and production of cheese in Chapter 9 Food manufacturing.

Uses of cheese

Cheese can be used in the making of both sweet and savoury dishes. It has many benefits and functions. Some of these are listed below.

- Cheese can provide flavour: when making a white sauce adding cheese gives improved flavour.
- Cheese can provide colour: when sprinkled on top of dishes and grilled or baked it will turn an attractive brown colour.
- Cheese can provide texture: when melted it can provide a soft, moist and stringy texture.
- Cheese can increase the nutritional value of a dish: when sprinkled on top of a baked potato it will provide additional nutrients such as protein, fat, calcium and vitamins.

Figure 1.13 **Broccoli with cheese sauce**

Storage of cheese

- Store all cheeses in a refrigerator between 0 and 5°C.
- Some soft cheeses will perish quickly and must be used within a few days.
- Hard cheeses like Cheddar can be stored for a longer period of time.
- Store cheese in an airtight box or package to prevent drying out.

Activity

1 Name four hard cheeses that are not included in the table.
2 Name two dishes that can be made using soft cheeses.
3 List three of the main nutrients found in cheese.

KEY WORDS

Fermented: when milk has been turned into a different food like cheese or yoghurt using harmless bacteria which enable the sugar to convert into lactic acid.

Ripened cheese: a cheese that has an external mould or a bloomy rind on the outside.

Unripened cheese: a cheese that has a soft and smooth texture and does not have a rind.

KEY POINTS: CHEESE

- Cheese is a solid or semi-solid form of milk.
- There are many different types and varieties of cheese available.
- Cheese contains HBV protein and is a good source of calcium.
- Cheese has many functions including providing flavour, colour, texture and nutritional value to a range of dishes.

Yoghurt

Yoghurt is made from milk. It is made by adding harmless edible bacteria to the milk, which causes it to ferment. This means the carbohydrate (sugar) in the milk, which is lactose, is converted into lactic acid by the bacteria. The lactic acid will set the milk's protein, which will thicken it. The lactic acid will also give the yoghurt its characteristically tangy flavour.

You will learn more about processing and production of yoghurt in Chapter 9 Food manufacturing.

Types of yoghurt

Yoghurt can be made from different types of milk. Some yoghurt will include additional ingredients such as sugar, which is used to sweeten it, or fruit and other flavours such as honey or vanilla.

Examples of types of yoghurt:

- **Set yoghurt:** is set in the pot in which it is sold. It has a firmer texture than other yoghurts.
- **Live yoghurt:** this yoghurt has been fermented with live culture bacteria which is still living.
- **Greek (strained) yoghurt:** is made from cow's or ewe's milk. It can be quite a thick yoghurt and is higher in fat.

Figure 1.14 Yoghurt

Nutritional value

Nutrient provided	Additional information
Protein	High Biological Value
Fat	Vanies according to the type of yoghurt. Some are made with whole milk which has a higher fat content, some are fat-free.
Calcium	A good source provided by the milk
Carbohydrates	In the form of lactose (sugar)
Vitamins	B and some A and D (and E if it is whole milk yoghurt)
Water	High water content

Table 1.8 Nutritional value of yoghurt

Storage of yoghurt

- Store in the refrigerator between 0°C and 5°C.
- Use within the use-by date.

KEY POINTS: YOGHURT
- Yoghurt is made using a harmless bacteria.
- Types of yoghurt include set, stirred, live.
- Yoghurt is a nutritional food as it has good calcium value and HBV protein.

Exam practice questions

1 Explain the differences between pasteurised milk and UHT milk. [6 marks]

2 A vegetable quiche can be made using cheddar cheese. Identify two different functions of the cheese and explain how the cheese fulfils each function. [4 marks]

3 Explain the importance of lactic acid in yoghurt making. [2 marks]

Meat

There are three animals that we generally use in the UK to provide us with meat:

- cows
- sheep
- pigs.

These animals provide a range of cuts of meat and a variety of different meat products for people to cook with. Meat is an important part of a daily diet for many people.

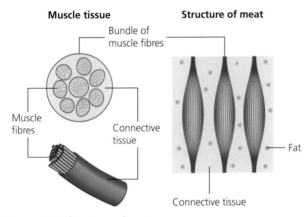

Muscle tissue   **Structure of meat**

Bundle of muscle fibres

Muscle fibres

Connective tissue

Fat

Connective tissue

Figure 1.15 **Structure of meat**

Figure 1.16 **Visible fat on sirloin steak meat**

Figure 1.17 **Different types of meat – beef, pork and lamb**

You will learn more about how meat is reared in Chapter 8 Food provenance.

Structure of meat

Meat is made up of:

- water
- protein
- fat.

Raw meat is muscle with **connective tissue** and fat. The muscles are bundles of fibres which are surrounded and held together by the connective tissue. These muscle fibres can be different lengths depending on which part of the animal they come from – if it is a part of the animal that does a lot work, for example, the neck or leg, the fibres will be longer; this can make the cut of meat tough. Therefore a suitable cooking method must be used to make the meat tender.

The fibres are very small tubes which contain water as well as mineral salts. The fat in meat is classed as visible or invisible.

- **Visible fat** can be seen around the edge of certain cuts like lamb and pork chops and in beef sirloin steaks.
- **Invisible fat** is found in the connective tissues of muscles and is referred to as marbling. Marbling will help to keep the meat moist during cooking.

Cows	Beef, veal	**Steaks** – sirloin, fillet, rump
		Joints – topside, brisket, silverside
		Cuts – skirt, chuck, minced
Sheep	Lamb, mutton	**Steaks** – shoulder, fillet
		Joints – leg joint, saddle, best end of neck
		Cuts – chump chops, loin chops, noisettes, mince
Pigs	Pork, bacon, gammon, ham	**Steaks** – shoulder, loin
		Joints – spare rib, leg, shoulder, loin
		Cuts – belly slices, spare ribs, chops

Table 1.9 **Types of meat**

Nutritional value

- Meat is a good source of high biological value protein.
- Fat amounts can vary.
- Meat provides fat-soluble vitamins A and D; amounts of B vitamins can vary.
- Meat provides a good source of iron.
- Meat has a high water content.

Cooking of meat

Meat is cooked for a variety of reasons:

- to make it safe to eat and kill harmful bacteria
- to make it tender to eat
- to improve the flavour and make the colour more appealing
- to make the nutrients more **digestible** and available to the body.

Cooking and storage methods

Meat is a high risk food that can contain harmful bacteria that could cause food poisoning, so it is important that it is cooked and stored correctly. When cooking, meat thermometers or temperature probes can be used to ensure the centre of a piece of meat has reached a safe temperature.

Raw meat should be stored:

- between 0–5°C in the refrigerator
- at the bottom of the refrigerator to avoid any juices dripping onto other foods
- in a covered container or on a covered plate and used within its use-by date if sold in a packet.
- If meat is stored frozen rather than chilled it must be in suitable packaging and kept at -18°C or below. It must be defrosted thoroughly before cooking.
- Thawed meat should never be re-frozen.

Cooked meat should be:

- cooled within 1.5 hours and then covered and refrigerated
- stored **above** raw meat in the refrigerator.

....

You will learn more about preparing and cooking meat in Chapter 11 Preparation and cooking techniques; and about correct storage in Chapter 7 Food spoilage.

....

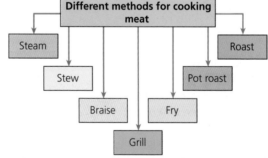

Figure 1.18 Different methods for cooking meat

KEY WORDS 🔑

Connective tissue: a protein called collagen found in meat.

Digestible: some foods are more quickly broken down by the digestive system than others. These foods are referred to as more digestible foods: food that can be broken down during the digestive process (by the action of enzymes) into the macronutrients or micronutrients that can be used by the body. These are absorbed by the body through the wall of the intestine.

KEY POINTS: MEAT 🎯

- Cows, sheep and pigs are three common animals providing meat in the UK.
- Meat is made up of protein, water and fat.
- Fat in meat is either visible or invisible.
- Meat provides HBV protein, iron, fat and some vitamins.
- Meat is a high risk food.
- Meat must be stored and cooked correctly to avoid food poisoning.

Activity

1 Name three different dishes that could be made using beef mince (ground beef).
2 Explain the difference between visible and invisible fat.
3 List two of the main nutrients found in meat.

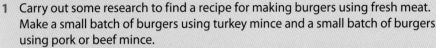

Practical activity

1 Carry out some research to find a recipe for making burgers using fresh meat. Make a small batch of burgers using turkey mince and a small batch of burgers using pork or beef mince.
2 To cook the burgers, bake one of each of the burgers in the oven and fry one of each of the burgers. Notice how much fat and juice has come out of the baked burgers and record some notes.
3 Carry out some taste testing with both samples of burgers and draw up a chart to record the differences in the taste, texture and appearance.
4 Carry out a nutritional analysis on each burger to find out if there are any differences in the nutritional value as well as the fat content.
5 Discuss your findings, stating which meat you would prefer to use for making burgers and why and which cooking method would be the most suitable.

Fish

Fish can be categorised into three main types:

- White fish – sole, halibut, cod, plaice, haddock, sea bass
- Oily fish – mackerel, salmon, trout, tuna, sardines
- Shellfish – crabs, lobsters, prawns, mussels.

Structure of fish

Fish is made up of:

- protein   ● water   ● minerals   ● fat

The flesh of fish is made up of muscle and connective tissue. The fish muscle has short fibres and the connective tissue which separates the muscles is very thin; this means fish can be cooked very quickly and still be very tender and moist.

Fresh fish can bought whole; it can also be cut into fillets, steaks and goujons. Frozen, smoked and canned are also others ways of buying fish.

You will learn more about how fish are caught in Chapter 8 Food provenance.

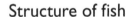

Figure 1.19 Common species of edible fish

Activity

1 Name three different white fish and three different oily fish.
2 Find out how many portions of fish we should eat each week and how much in weight one portion of fish is.
3 Cod is a popular white fish. Name three different dishes that could be made using this fish.

Nutritional value

The nutritional value of fish will vary depending on the type of fish. Generally fish is:

- high in protein – HBV
- low in fat – white fish and shellfish (although some fat is provided in the form of unsaturated oils)
- a good source of essential fatty acids – oily fish
- a good source of vitamins A and D – oily fish, for example, herring
- high in calcium if the bones are eaten – canned sardines and pilchards.
- shellfish may contain high levels of cholesterol.

Storage and cooking of fish

All types of fresh fish spoil very quickly, which results in them being unsafe to eat. They should be:

- used as soon as possible, preferably on the same day they were bought

- stored in a refrigerator
- cooked at a high enough temperature for a suitable amount of time to destroy any bacteria present.

Fish can cooked in a variety of ways to make it more interesting; it can be served with a variety of other foods, for example parsley sauce or on a bed of spicy noodles to increase the nutritional value of the dish and to add more flavour.

You will learn more about preparing and cooking fish in Chapter 11 Preparation and cooking techniques; and about correct storage in Chapter 7 Food spoilage.

KEY POINTS: FISH

- The three main types of fish are white fish, oily fish and shellfish.
- Fish can be cooked quickly and still be moist and tender.
- Fish provides HBV protein, some fat, calcium if bones are eaten and vitamins A and D.
- Fish spoils quickly; it must be stored correctly and eaten as soon as possible.

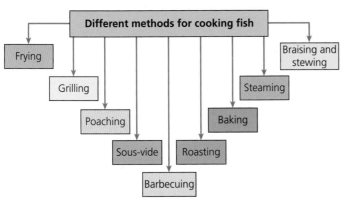

Figure 1.20 Different cooking methods for fish

Poultry

Chicken, turkey, duck, goose, guinea fowl and pigeon are all birds. These birds are classed as poultry. They are **reared** for their meat and sometimes eggs.

- Chicken is the most popular poultry used in the UK. Chickens are sold as a whole bird, or jointed as breast, legs, thighs and wings.
- Turkey is a larger bird and very similar to chicken. Many people associate turkey with Christmas time.
- Duck and goose are richer tasting birds; the flesh is fatty compared to chicken and turkey.

You will learn more about how poultry is reared in Chapter 8 Food provenance.

Chicken is a versatile food that can be combined well with other ingredients to make many different dishes, it is suitable for:

- frying
- roasting
- stewing
- boiling
- grilling

Structure of poultry

Poultry is very similar in structure to other meats, so it has muscle fibres and connective tissue. The breast of poultry is softer than the legs, which can be tough depending on the amount of movement the muscles have had. Older birds are not as tender as younger birds.

Nutritional value

The nutritional value of poultry will vary depending on the type of bird, the age of the bird, how it is reared and the part of the animal that is eaten.

Figure 1.21 Chicken

Figure 1.22 Turkey

- Poultry is a good source of HBV protein.
- Poultry provides fat (saturated fat – less than that found in meat).
- Poultry provides a good source of B vitamins.
- Poultry provides some vitamins A and D.

Storage and cooking of poultry

Poultry is a high risk food; this means it is high in protein and moisture. It can carry bacteria which can lead to **food poisoning** such as salmonella. It must be:

- stored in the coolest part of the refrigerator at the bottom
- stored and covered correctly to prevent any drips going onto other foods
- used within its use-by date
- cooked thoroughly both inside and outside at a temperature high enough to kill bacteria
- thawed thoroughly after freezing before cooking.

You will learn more about preparing and cooking poultry in Chapter 11 Preparation and cooking techniques; and about correct storage in Chapter 7 Food spoilage.

Activity

1 Name two examples of poultry.
2 Explain why the breast of chicken is considered to be a better portion than a leg.
3 Name two different ways duck can be served as a dish.

KEY WORDS

Reared: the conditions an animal is raised in when young.

Food poisoning: illness caused by bacteria or other toxins in food.

KEY POINTS: POULTRY

- Poultry includes chicken, turkey, goose, duck, guinea fowl and pigeon.
- Poultry is very similar to meat in structure.
- Chicken and turkey have less fat than other meats.
- Poultry must be stored correctly to prevent contamination of other foods.
- Poultry is a high risk food which can cause food poisoning.

Eggs

There are many different types of eggs available in the UK produced by hens, ducks, quails and geese. The most popular eggs we consume and use in food preparation and cooking are hen eggs.

Eggs can be cooked by a range of methods; an example of this would be at breakfast time when you can choose eggs that have been:

- boiled
- poached
- fried
- scrambled.

Eggs have many functional properties; this means they carry out important jobs when making many different food products, for example, cakes, sauces, desserts.

You will learn more about functional properties of eggs in Chapter 6 The effect of cooking on food and Chapter 11 Preparation and cooking techniques.

The table below shows the different sizes of eggs sold in shops. Many supermarkets have started to sell boxes of eggs that are of mixed sizes.

Egg size	Weight of egg
Small	53 g or under
Medium	Between 53 g and 63 g
Large	Between 63 g and 73 g
Very large	73 g and over

Table 1.10 Egg sizes

Figure 1.23 Enriched cage egg production

Some of the types of hen eggs available in shops include:

- Enriched cage farm eggs – these have replaced battery cages. Hens are kept in cages in sheds to promote egg laying. They have more space and height than in battery cages, a resting area and perches for hens.
- Free range eggs – this type of farming allows hens to walk around outside, scratch the soil and peck for food, sit up on perches and lay eggs in nests.
- Barn eggs allow the hens to roam freely inside uncaged, and have perches to roost on.

Eggs have a **porous** outer shell. The shell can be brown or white and makes up 10 per cent of an egg's structure. Inside the shell there are two parts:

- egg white: this amounts to 60 per cent
- egg yolk: this amounts to 30 per cent.

Nutritional value

Eggs will provide a range of nutrients:

- protein – in both the white and yolk
- vitamins – A, D and E in the yolk; B in the white
- iron – in the yolk
- fat – in the yolk
- water – in both the white and yolk.

Storage of eggs

- Eggs should be stored pointed end down in the refrigerator.
- They should be kept away from strong-smelling foods as the egg shell is porous and allows odours to be absorbed.
- Best-before dates should be noted to ensure the use of eggs when they are at their best.

Figure 1.24 **Free range hens**

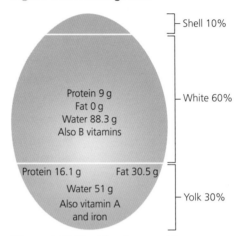

Shell 10%

Protein 9 g
Fat 0 g
Water 88.3 g
Also B vitamins

White 60%

Protein 16.1 g   Fat 30.5 g
Water 51 g
Also vitamin A and iron

Yolk 30%

Figure 1.25 **Nutritional composition of eggs (nutrients per 100 g)**

KEY WORDS

Porous: contains tiny holes which allows moisture and gases to pass through.

KEY POINTS: EGGS

- Hen eggs are mainly used in food preparation and cooking.
- Eggs can be bought in four different sizes.
- Egg types depend upon the conditions the hens have been kept in.
- The biggest part of the egg is the white.
- Eggs provide a range of nutrients including protein, fat-soluble vitamins and iron.
- Eggs should be stored in the refrigerator, away from strong smelling foods.

Practical activity

Cook three eggs by frying, boiling and scrambling. Compare the taste, texture and appearance and note your findings. Find out how the different methods of cooking affect the nutritional value of the eggs.

Activity

1 Carry out some research on the types of eggs sold in supermarkets.

   a) Compare the costs of the different eggs.

   b) Explain which eggs you would buy and why.

Exam practice questions

1 Explain why fish takes less time to cook than meat. [2 marks]
2 Cod fillet can be cooked by different methods like frying and baking. Explain which method of cooking would be healthier and why. [3 marks]
3 Give **two** detailed reasons for cooking meat. [4 marks]
4 Explain why some people may refuse to buy battery farm eggs. [4 marks]

Soya, tofu, beans, nuts, seeds

Soya

Soya comes from soya bean pods that are located in the soya plant. Soya beans are part of the legume family. Beans, peas and lentils are also part of this family.

Soya beans are usually green but they can be yellow, brown or black. Young soya beans are called edamame beans; these can eaten fresh from the pod and are sometimes included in pre-made chilled salads.

The texture of a soya bean is very adaptable; this allows the beans to be processed into a variety of foods such as:

- soya milk – a plant milk produced by soaking dried soya beans and grinding them in water.
- soy sauce – made from a fermented paste of boiled soya beans and other ingredients which is pressed to produce a liquid.
- miso – a fermented soya bean paste used as a flavouring in Asian cooking.
- soya flour– made by grinding whole dry soya beans into flour.
- tempeh – made from cooked and slightly fermented soya beans and formed into a patty, similar to a very firm veggie burger.

Many supermarkets and health food shops sell other soya products such as desserts, yoghurts and margarine.

Figure 1.26 A range of soya products – soya beans, chunks, and tofu

Soya beans can be bought:
- dried
- canned
- fresh – as edamame beans.

Nutritional value

- Soya beans are a good source of high biological value protein.
- Soya beans can be rich in calcium and magnesium.
- Soya mince can contain a good amount of fibre.

Storage of soya products

The storage of soya products depends upon the type of product; chilled products are stored in refrigerators. Cupboard ingredients should be stored in cool, dry places. For soya beans:

- Dried and canned soya beans should be stored in a cool, dry place until cooked or opened and then stored in an airtight container in the refrigerator for up to three days.
- Fresh edamame beans should be stored in the refrigerator.

Tofu

Tofu is sometimes referred to as bean curd. It is made from fresh soya milk that has been curdled; it is then pressed into a solid block and cooled.

This is the same process that is used to make traditional dairy cheese. The liquid, which is referred to as whey, is removed and the curds are pressed to form a firm block.

Tofu is a bland-tasting food, so it must be cooked with other stronger flavoured foods so it can absorb those flavours. It can be cooked in different ways, which can change its texture from quite smooth and spongy to more crisp and crunchy.

Tofu can be used to make a number of dishes such as:

- stir fries
- curries
- salads.

Nutritional value

The nutritional value of tofu is similar to soya:

- It is a good source of high biological value protein.
- It contains iron, calcium and other minerals.
- It provides some B vitamins.

Buying and storing tofu

- Tofu can be bought as a chilled product which must be stored in a refrigerator.
- It is also sold in sealed containers. The tofu must be rinsed when opened and stored in water in a container in the refrigerator.
- Tofu can be frozen if in its original packet and unopened.

KEY POINTS: SOYA AND TOFU

- Soya beans can be processed into many different products such as soya milk, soya flour, soy sauce.
- Soya products such as tempeh, tofu and soya mince are eaten as alternatives to meat as well as for their nutritional qualities.
- Soya and tofu can have a bland flavour so they must be mixed or cooked with other stronger flavoured ingredients to allow them to absorb flavour.

Activity

Research the shops and supermarkets in the area in which you live to find out if they sell any soya products. Find out which soya products they sell and present your findings in a chart.

Practical activity

1 Following the same recipe, make two small samples of bolognaise: one using beef mince and one using, soya mince.
2 Ask people to taste test them and note their comments.
3 Compare the nutritional differences between the two samples using a nutrition software package. Which sample was more in line with recommended healthy eating guidelines?

Figure 1.27 Different types of beans

Figure 1.28 Mixed bean salad

Beans

Many people if asked to name a bean would say 'baked beans'. This type of bean is very popular and sold in large quantities in the UK. Baked beans are haricot beans, which are also known as navy beans, and are sold in a tomato sauce.

Beans are classed as legumes, but are generally referred to as pulses. Pulses are edible seeds that grow in pods. There are many different types of beans.

Beans are sold in shops in different forms, most supermarkets will sell them:

- fresh
- frozen
- dried
- canned.

Beans are good store cupboard ingredients; they can be added to many different dishes to add colour, bulk, a different texture, flavour and improve the nutritional value of a dish.

Nutritional value

Most beans will provide:

- some carbohydrates
- a high amount of protein
- some B group vitamins
- iron and calcium
- a good amount of dietary fibre.

Storage of beans

This will depend upon the type of bean:

- Fresh beans should be stored in the refrigerator and used within the use-by date.
- Frozen beans should be stored in the freezer.
- Dried beans should be stored in airtight containers once opened and in a cool, dry place – after soaking they should be covered and stored in the refrigerator.
- Canned beans should be stored in a cool, dry place. Once opened they should be stored in an airtight container and refrigerated.

KEY POINTS: BEANS

- Beans are called legumes as well as being referred to as pulses.
- Beans are sold in many different forms and can be added to many savoury dishes to improve the nutritional value.
- Beans provide a good source of protein and fibre.

Nuts

When asked to describe what a nut is, it can be quite difficult to be precise. Some nuts are edible kernels from which the fruit wall has been removed (almonds, walnuts). Some are seeds (brazil, pistachio and pine nuts) some are pulses (peanuts) and some are fruits with a dry shell (cashew, chestnuts and hazelnuts).

There are many different types of nuts available that can be used in the cooking of savoury dishes, for example a nut roast and chicken satay. Nuts can also be used in baking cakes, biscuits or sweet treats, or just simply eaten as a healthy snack.

Some adults and children can have an allergy to nuts. Peanuts are the most common nuts that cause allergic reactions. A severe reaction to nuts is called anaphylaxis and can be life-threatening. Symptoms of a severe nut allergy can include breathing difficulties and swelling of the throat.

Some example nuts include:

- hazelnuts
- chestnuts
- almonds
- pecan nuts.
- walnuts
- pistachio nuts
- brazil nuts

Peanuts can sometimes be called a groundnut, but they are not a true nut – they are actually a type of pea that grows underground which makes them a legume.

For more information on nut allergies see Chapter 4 Plan balanced diets.

There are many different ways of buying nuts. Some of these include:

- blanched
- whole
- flaked
- roasted
- shelled
- chopped
- ground
- salted.

Nutritional value

- Energy values can be quite high due to the high fat content of some nuts.
- Low biological value protein.
- B group vitamins are found in all nuts.
- Calcium and iron amounts can vary depending on the type of nut.
- Nuts will provide dietary fibre.

Storage of nuts

- If not stored correctly nuts can become rancid due to the oil content.
- Nuts should be stored in airtight containers to keep out moisture and strong odours.
- It is recommended to use nuts by the best-before date.

Figure 1.29 Different types of shelled nuts

KEY POINTS: NUTS

- When defining nuts, some are edible kernels from which the fruit wall has been removed, some are seeds and some are pulses.
- Nuts can be used in the cooking of savoury dishes as well as sweet products.
- Nuts can cause severe allergic reactions.
- Nuts can be brought in many different forms such as shelled, ground, chopped.
- Nuts can have high energy values, as well as providing dietary fibre, protein and B vitamins.

Seeds

Seeds come in a variety of shapes, colours and sizes. Most supermarkets will offer a range of seeds to be used in the preparation and eating of food; specialist shops tend to offer a more varied selection.

Figure 1.30 A mix of seeds: sunflower, pumpkin and linseed

Sunflower seeds are one of the most commonly available seeds. Other types of seeds include:

- black onion
- poppy
- pumpkin
- linseed
- fennel
- sesame.

Seeds have many uses, some of these include:

- mixed and eaten as a healthy snack
- used as a topping on foods
- roasted or toasted and added to salads to add texture and additional nutrients
- ground and mixed with foods to add flavour
- used in the manufacture of oils.

Nutritional value

Seeds can provide a range of nutrients. Some of these include:

- protein
- essential fatty acids
- essential minerals such as iron and zinc
- Vitamins B and E.

Storage of seeds

Seeds are best stored in airtight containers in cool, dry places. They will also be provided with a best-before date.

KEY POINTS: SEEDS
- Popular seeds include poppy, pumpkin and sunflower.
- Seeds have many uses and can be included in many different food products to increase nutritional values.
- Many seeds are used in the manufacture of oils.

Exam practice questions

1 Explain what the term 'legume' means. [2 marks]
2 Name **three** different products that can be made using soya beans. [3 marks]
3 Tofu has a bland flavour. Explain how you could improve this when using it to prepare dishes. [2 marks]
4 Explain why nuts can be a health risk to certain people. [3 marks]

Butter, oils, margarine

Butter

Butter is made from cream that is churned or moved around quickly until lumps of butter form. It is a solid, firm fat. When warmed or heated it will change from a solid to a soft consistency and melt. There are different types of butter sold in shops; the two main ones are:

- salted butter
- unsalted butter.

When preparing different dishes butter has many uses. Some of these include:

- Melting gently and pouring over cooked vegetables to add flavour and improve appearance.
- Spreading on foods like bread and crackers to provide flavour and moistness.
- Creaming with sugar when making cakes to provide **aeration** and give a richer flavour.
- Shallow-frying different foods (e.g. an omelette) to add flavour and prevent sticking.
- **Shortening**: when making pastry, butter is rubbed into flour to give a short, crumbly texture.

Nutritional value

- Butter is high in fat which is mostly saturated.
- Butter provides vitamins A and D.
- Salted butter provides sodium (salt).

Storage of butter

- Butter should be kept in a refrigerator; it is suitable for freezing.
- Butter should be kept away from light and strong-smelling foods.
- Butter should be kept fully covered or stored in a lidded container as it can be affected by air, which can make it go rancid.

Oils

Oils are liquid at room temperature. They can be lighter than a solid fat such as butter, and easier to digest. Vegetable oils are natural oils found in seeds, nuts and some fruit.

Some examples of vegetable oils include:

- sunflower oil
- olive oil
- sesame oil
- corn oil
- rapeseed oil.

General uses of oils include:

- frying
- basting
- marinating
- dressings.

Some oils are more suitable for particular recipes and uses such as drizzling or sprinkling over food due to their flavour. Flavoured oils are also very popular (e.g. chilli oil).

KEY WORDS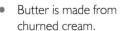

Aeration: the adding of air during the combining of different ingredients.

Shortening: when flour is coated with fat to prevent gluten formation.

KEY POINTS: BUTTER

- Butter is made from churned cream.
- Butter is a solid, firm fat that should be kept in a refrigerator.
- Butter has many uses when preparing dishes such as spreading and shallow-frying.
- Butter is high in saturated fat.
- Butter must be stored away from light and strong-smelling foods. Correct storage can prevent it from going rancid.

Figure 1.31 **Butter**

Figure 1.32 Oils

Nutritional value

The main nutrient provided by oil is fat; this is unsaturated. This means the fat comes from a vegetable, nut or seed, rather than an animal source. This type of fat is considered as a 'healthy' fat.

Storage of oils

Oils should be stored in a cool, dark place away from direct sunlight.

> ### KEY POINTS: OILS
> - Oils are liquids.
> - Vegetable oils are naturally found in some fruits, seeds and nuts.
> - Oils are high in unsaturated fats.

Margarine

Margarine was introduced as an inexpensive alternative for butter. It is made from vegetable oils and has vitamins A and D added to it by law.

There are many different brands of margarine produced by different manufacturers, giving consumers a wide choice. These can vary in flavour. Margarine is sold either in blocks as a hard margarine or in tubs as a soft margarine.

Uses of margarine:
- Block margarine can be used for baking, it is recommended for making pastry, and it can also be used for frying foods.
- Soft margarine can be used for baking; some cake recipes recommend using this rather than butter. It can also be used as a spread when making sandwiches. It is not suitable for frying.
- Some soft margarines have a very low fat content making them unsuitable for making cakes, biscuits and pastry.

Nutritional value

This will depend on the type of margarine, but most will provide:
- a high amount of fat
- some water
- vitamins A and D
- minerals such as sodium.

Storage of margarine

Margarine should be stored in the refrigerator and used within its use-by date.

Activity

1 Name three oils that are not included in the list.
2 Explain why oils should not be stored in the fridge.
3 Name three ingredients that could be used to flavour oils.

KEY POINTS: MARGARINE

- Margarine is made from vegetable oils.
- Margarine comes in two forms - hard and soft.
- Margarine has vitamins A and D added to it by law.

Exam practice questions

1 Explain the functions of butter when using it to make a shortcrust pastry. [4 marks]
2 Give **two** differences between butter and margarine. [2 marks]
3 Suggest **three** different uses for olive oil when preparing and cooking foods. [3 marks]

WJEC EDUQAS GCSE Food Preparation and Nutrition

Sugar and syrup

Sugar

Sugar comes from **sugar cane** or **sugar beet**. Sugar is pure carbohydrate; it is referred to as providing 'empty calories' as it does not provide any other nutrients.

There are many different types of sugar available. Some of the most popular types are included in the table below.

Sugar type	Description	Uses
Granulated	White sugar, quite coarsely ground small crystals	For sweetening drinks like tea, sprinkled on cereals, can be used in some sweet baked goods
Caster	White sugar, made from ground granulated sugar, a fine crystal	Used in cake making, e.g. Victoria sandwich
Icing	White sugar, made from ground granulated sugar, a fine powder	For cake decorating and icing, e.g. buttercream
Demerara	Pale brown, made from raw sugar, has larger, more coarse crystals than granulated	For dishes that like extra crunchiness, e.g. baked apples, flapjacks, crumble toppings
Soft brown/dark brown	Brown sugars, small sugar crystals that contain molasses, a dark syrup	For colour in cakes, e.g. Christmas cake

Table 1.11 **Types of sugar**

Sugar performs many functions in the manufacturing of different food products. One of its primary functions is to act as a sweetener for both sweet and savoury products.

You will learn more about the functional properties of sugar in Chapter 6 **The effect of cooking on food.**

Storage of sugar

Sugar should be stored in a cool, dry place away from moisture, heat, and strong flavours and odours. Sugar has a long shelf life if stored correctly.

Syrup

Golden syrup is the most familiar syrup to many people. It can be bought in various styles including the traditional tin, in a squeezy bottle to make pouring easier when using it for cooking, and in a flip cap breakfast bottle that does not drip.

Golden syrup is gold in colour and is very sweet. It can be used in the making of both sweet and savoury products such as baked goods, marinades for meat, salad dressings and desserts.

Black treacle is also classed as a syrup. It has a much darker brown colour than golden syrup and is thicker in consistency. It also has a much stronger flavour. Black treacle is used in the making of Christmas puddings, gingerbread and some curry sauces.

Figure 1.33 **Different types of sugar**

Figure 1.34 **Golden syrup**

Storage of syrups

Syrups are best stored in a cool, dry cupboard and are best used within three months of opening.

KEY WORDS

Sugar cane: a tall grass grown in hot countries.

Sugar beet: a root crop similar to a parsnip grown in a field in countries with warm and cold seasons.

KEY POINTS: SUGAR AND SYRUP

- Sugar is pure carbohydrate and does not provide any other nutrients.
- There are many different types of sugar available.
- Sugar is used in the making of both sweet and savoury dishes.
- Golden syrup and black treacle are sweet syrups.

Exam practice questions

1 Describe the differences between granulated and icing sugar. [2 marks]

2 Explain why we are advised to reduce the amount of sugar we consume. [4 marks]

SECTION 2

Principles of nutrition

This section contains the following chapters:

CHAPTER 2
Macronutrients and micronutrients

Learning objectives

In this chapter you will learn about:
- the differences between macronutrients and micronutrients
- the roles or functions of macronutrients and micronutrients in our diet and our bodies
- where we can find macronutrients and micronutrients in the food we eat
- how these nutrients work together in our bodies
- what happens if we lack nutrients.

Our bodies need fuel and chemicals to grow, maintain and repair all the cells and organs, and to make them work properly. Imagine if you put the wrong fuel into a car – it would not work properly. The same rules apply to our bodies. We have to eat food that contains everything we need to provide the energy and chemicals our bodies require to keep them as healthy as possible.

The fuel and chemicals that we need for our bodies are called **nutrients**.

The nutrients that we eat are divided into two main groups:
- **Macronutrients** (*macro* means large): these are needed by the body in large amounts. These nutrients are **proteins**, **fats** and **carbohydrates**.
- **Micronutrients** (*micro* means small): these are needed by the body in small amounts. These nutrients are **vitamins, minerals** and **trace elements**.

The body also needs other substances in food to work properly, including **water** and **fibre**.

Protein

Protein is a macronutrient and is needed by the body for the following reasons:
- It provides all the chemicals to make the body **grow**, particularly in children and pregnant women.
- It provides all the chemicals to help the body **repair** any damage after illness, accidents and surgery.
- It **maintains** the body to keep it working well, producing **enzymes** for digestion, muscle activity, nerve function and **hormones**, which regulate some body functions.
- It provides a secondary source of **energy** for the body.

Proteins are very large molecules and are made of small units called **amino acids**. There are many different amino acids that are joined together in different ways and different numbers to produce different proteins.

Some amino acids are known as **essential amino acids**. These are the amino acids that cannot be made by our bodies, so we must eat the proteins that contain them. There are nine essential amino acids needed by our bodies. These are:

- histidine
- isoleucine
- leucine
- lysine
- methionine
- phenylanine
- threonine
- tryptophan
- valine.

All the other amino acids can be made by our bodies from the protein we eat. These are the **non-essential amino acids**. There are eleven of these. These are:

- alanine
- arginine
- asparagine
- aspartic acid
- cystenine
- glutamic acid
- glutamine
- glycine
- proline
- serine
- tyrosine.

High Biological Value (HBV) proteins

Foods that contain all of the essential amino acids are called High Biological Value proteins (HBV proteins).

Low Biological Value (LBV) proteins

Foods that contain only some of the essential amino acids are called **Low Biological Value proteins (LBV proteins)**. You will have to eat a mixture of these foods every day to get all of the amino acids you need.

Which foods contain protein?

The following foods are sources of HBV protein:

- meat
- fish
- eggs
- milk
- cheese
- soya beans.

Figure 2.1 These foods are sources of HBV protein

Practical activity

You can use soya products as a vegetarian or vegan source of HBV protein.

Plan a meal for a vegetarian, using a soya product as a source of HBV protein. Work out how many grams of protein are in the meal and check if it is correct for the age group, using Table 2.1 giving the Dietary Reference Values (DRVs).

Activity

Match the words below with their description.

LBV proteins	Needed by the body in large amounts.
Amino acids	Contain all the amino acids needed by the body.
HBV proteins	The units from which proteins molecules are made.
Macronutrients	Contain some of the amino acids needed by the body.

Other protein foods are sources of LBV protein. These include:

- cereals (rice, oats, quinoa, wheat, millet)
- peas, beans (except soya beans) and lentils
- nuts and seeds.

If we combine LBV proteins in a meal, we can provide all the essential amino acids for our bodies. This is called **food combining** or using **complementary proteins**. An example of a meal using complementary proteins is beans on toast or rice and dhal.

What happens if we eat too little or too much protein?

Because protein is such an important macronutrient, either too much or too little will have a big impact on the body.

If children have too little protein in their diet, they:

- stop growing, or grow slowly
- may have thinning hair, or hair loss
- may experience a change in skin colour and become paler
- cannot digest food properly, and may have diarrhoea
- catch infections, like colds, easily
- have low energy levels
- lose weight and become thin and weak
- may have a build-up of fluid under their skin (this is called oedema).

If you are an adult, too little protein will have the following effects:

- You may lose fat and muscle from your body.
- Fluid may build up under the skin (oedema).
- You may lose weight.
- Your cuts and bruises may be slow to heal.
- You lack energy.
- Your hair and skin becomes dry.
- You catch infections more easily.

Too much protein will also have an effect on your body:

- Protein is processed by the kidneys and liver, so too much protein will put a strain on these organs.
- You may put on weight, as extra protein is converted into fat which is then stored in the body.

How much protein do we need?

Everyone needs to eat foods which contain protein every day, but how much protein we need depends on our age, our lifestyle and our activities.

Babies, children and teenagers are still growing and therefore need more protein for this, as well doing all the other things in their bodies that require protein.

Adults still need protein to help their hair and fingernails grow and for the body to repair.

Figure 2.2 These are LBV sources of protein

Stretch & challenge

Write out all the foods you ate yesterday. Identify all the proteins you ate and say whether they are HBV or LBV sources.

Age	DRVs (in grams) for protein	
	Male	Female
0–3 months	12.5	12.5
4–6 months	12.7	12.7
7–9 months	13.7	13.7
10–12 months	14.9	14.9
1–3 years	14.5	14.5
4–6 years	19.7	19.7
7–10 years	28.3	28.3
11–14 years	42.1	41.2
15–18 years	55.2	45.0
19–50 years	55.5	46.5
50+ years	53.3	46.5
During pregnancy		Add another 6 grams
During lactation		Add another 11 grams

Table 2.1 Dietary Reference Values for protein

Pregnant women need protein to allow their baby to develop, and women who are breastfeeding (lactating) need protein to make their milk.

Nutritionists and scientists have worked out how much protein is needed by individuals. These are called the **Dietary Reference Values (DRVs)**.

Practical activity

Plan and prepare a main course that uses LBV proteins as complementary proteins. Work out the total protein content per portion and see if it is suitable for a teenager.

KEY WORDS

Macronutrients: nutrients needed by the body in large amounts. They include proteins, fats and carbohydrates.

Micronutrients: nutrients needed by the body in small amounts. They include vitamins, minerals and trace elements.

Amino acids: small units that join together to make large molecules of proteins.

Essential amino acids: the nine amino acids that cannot be made by our bodies, so we must eat the proteins that contain them.

High Biological Value (HBV) proteins: foods that contain all the essential amino acids.

Low Biological Value (LBV) proteins: foods that contain some of the essential amino acids.

Complementary proteins: LBV proteins that are eaten in one meal together to provide the essential amino acids.

Dietary Reference Value (DRV): the amount of a nutrient a person needs.

KEY POINTS: PROTEIN

- Protein is a macronutrient.
- Protein is needed for growth, repair and maintenance of the body, to make enzymes and hormones, and as a secondary energy source.
- Proteins are made of amino acids.
- There are nine essential amino acids. Foods containing these are known as HBV proteins.
- HBV proteins are found in meat, fish, eggs, milk, cheese and soya.
- LBV sources of protein are cereals, peas, beans, lentils, nuts and seeds.
- If we do not have enough protein we will become thin and weak.
- Babies, children, teenagers, pregnant women and lactating mothers need more protein than adults.

Exam practice questions

1 List **three** functions of proteins in the diet.   [3 marks]
2 Explain the difference between HBV and LBV proteins.   [4 marks]
3 Name **three** sources of HBV protein.   [3 marks]
4 List **four** symptoms you might see in a child who is not getting enough protein in their diet.   [4 marks]
5 What happens if we have too much protein in our diet?   [3 marks]

Fats and oils

Fat is a macronutrient. A general term for fats is **lipids**.

Fats can be either solid or liquid at room temperature. Oils are liquid at room temperature.

Fat is necessary in the diet for the following reasons:

● Fats are a good energy source, and also form the structure of some cells.
● Fat is stored under the skin where it helps insulate the body against the cold.
● Our vital organs, such as the liver and kidneys, are protected by a layer of fat.
● Fat is a good source of vitamins D, E and K.
● Fat gives food texture and flavour.
● Fat in our diet helps give a feeling of **satiety** (feeling full after eating).

Fats are made of combinations of carbon, hydrogen and oxygen atoms. They are composed of fatty acids and glycerol. There are differences in how these are combined in the different types of fats.

Typical values	Per 100 g	Per 40 g bar
Energy	1181 kJ	724 kJ
	432 kcal	173 kcal
Protein	6.1 g	2.4 g
Carbohydrates:	59.4 g	23.8 g
Of which sugars	29.9 g	11.9 g
Fat:	18.8 g	7.5 g
Of which saturated fat	9.5 g	3.8 g
Sodium	0.06 g	0.02 g
Salt	0.15 g	0.06 g

Figure 2.3 Saturated fatty acids are full up with hydrogen atoms

Saturated fats

In **saturated fats**, each carbon atom is combined with two hydrogen atoms.

Most saturated fats are solid at room temperature. Examples of saturated fats are butter, lard, suet and animal fat found on meat.

Too much saturated fat in the diet has been linked to high blood cholesterol, an increased risk of heart disease, diabetes and obesity.

Unsaturated fats

There are two types of **unsaturated fats**: **monounsaturated** and **polyunsaturated**. Unsaturated fats are usually liquid or soft at room temperature and have a lower melting point.

Monounsaturated fats contain a pair of carbon atoms with only one hydrogen atom attached, so they are capable of taking another carbon atom. They are soft at room temperature, but will harden when put in the fridge. They are found in animal and vegetable fats and are considered healthier because they can help to lower blood cholesterol, reduce the risk of diabetes and are linked with a lower rate of cancer.

Figure 2.4 Sources of saturated fat in the diet

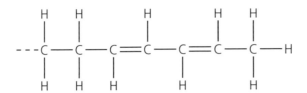

Figure 2.5 Unsaturated fatty acids are not full up with hydrogen atoms – they could take more by breaking the double bonds and adding hydrogen

Figure 2.6 Sources of monounsaturated fat

WJEC EDUQAS GCSE Food Preparation and Nutrition

Polyunsaturated fats have two or more pairs of carbon atoms, which are capable of taking up more hydrogen atoms. They are soft and oily at room temperature and will not harden in the fridge.

Manufacturers sometimes change the structure of fats in products by adding hydrogen to vegetable oils. This creates manmade molecules called **trans-fatty acids**. This process is called **hydrogenation**. It turns oils into solid fats, and is much cheaper to use in products than those normally used, such as solid fats like butter – for example, in cakes and biscuits. It makes them behave like saturated fats.

Recently, medical research has found that these trans-fatty acids are very bad for your cardiovascular system, and may increase your risk of heart disease and breast cancer. Many manufacturers are therefore now reducing the use of these manmade molecules.

Figure 2.7 Sources of polyunsaturated fat

Activity

Match the words below with the correct definition.

Saturated	The process of turning oils into solid fats.
Unsaturated	A fat which is usually solid at room temperature. Examples include butter, lard and suet.
Monounsaturated	A fat that has two or more pairs of carbon atoms which are capable of taking up more hydrogen atoms. They are soft and oily at room temperature and will not harden in the fridge.
Polyunsaturated	A fat containing a pair of carbon atoms with only one hydrogen atom attached. They are soft at room temperature, but will harden when put in the fridge.
Hydrogenation	A fat with a lower melting point that is usually liquid at room temperature.

Stretch & challenge

Describe, in detail, what a fat is and what an oil is. Use a diagram to explain the chemical differences.

Sources of fat

Fats come from animal and plant sources.

Animal sources include:
- meat and meat products
- dairy products, e.g. milk, cheese, butter and cream
- fish, particularly oily fish like tuna, sardines and salmon.

Plant sources include:
- avocados and olives
- nuts and pulses, e.g. peanuts and walnuts
- seeds such as sesame, sunflower and soya.

Figure 2.8 Visible fats and oils

Figure 2.9 Invisible fats

Visible and invisible fats

Some fats are **visible**, such as the fat on meat, or in the butter or oils that we use for frying or salad dressing.

Other fats are **invisible** and form part of a product that we eat, such as biscuits, ice cream or ready meals.

Essential Fatty Acids (EFAs)

Essential fatty acids are fatty acids that cannot be made by the body, but are important for your body to make it function efficiently. The balance of EFAs is important for us, as they are essential for regulating body processes, including blood clotting and control of inflammation.

Two important ones are:

● **Omega 3**, found in oily fish, seeds, walnut oil and green leafy vegetables. It helps protect the heart.
● **Omega 6**, found in vegetables, fruits, grains, chicken and seeds. It helps lower blood cholesterol.

What happens if we eat too much or too little fat?

Too much fat

Fat is a high energy source, providing over twice the amount of energy as carbohydrates. If we do not use up the energy from the fat we consume, it will be stored as fat in our body and we will gain weight. If we eat too much fat, the extra fat may be stored in our liver and cause health problems. It can also increase the risk of stroke.

Eating food high in saturated fat can raise blood cholesterol levels and increase the chance of heart disease.

Hydrogenated fats can increase the risk of cancer, diabetes, obesity and bone problems.

Too little fat

Some fatty acids are essential for the correct growth and functioning of the body. If babies and children lack these their normal growth will be affected. If we do not get enough energy from fat or carbohydrate, we will use up our fat stores and become thinner. We may also feel colder.

How much fat should we eat per day?

Most people eat too much saturated fat. A gram of fat provides 9 kcal, compared with 4 kcal per gram of carbohydrate.

● The average man should not eat more than 95 g of fat per day, of which not more than 30 g should be saturated fat.
● The average woman should not eat more than 70 g of fat per day, of which not more than 20 g should be saturated fat.
● A child's diet should aim to have about 35 per cent of total intake of food as fat. This ensures the child is getting sufficient energy sources and vitamins D, E and K.

Practical activity

Make a full-fat and a reduced-fat version of the same dish, using either lower fat products or a reduced-fat substitute. Test them out on a taste panel and see if the panel can identify which is which.

WJEC EDUQAS GCSE Food Preparation and Nutrition

KEY WORDS

Lipids: a general term given to fats.

Satiety: feeling full after eating.

Saturated fats: fats with two hydrogen atoms for each carbon atom that are mainly solid at room temperature and mainly animal fats.

Unsaturated fats: fats which are usually liquid or soft at room temperature.

Monounsaturated fats: fats containing a pair of carbon atoms with only one hydrogen atom attached. They are soft at room temperature, but will harden when put in the fridge. They are considered to be healthier than other fats.

Polyunsaturated fats: have two or more pairs of carbon atoms which are capable of taking up more hydrogen atoms. They are soft and oily at room temperature and will not harden in the fridge.

Trans-fatty acids: manmade molecules created when manufacturers add hydrogen to vegetable oils (hydrogenation).

Hydrogenation: the process of turning oils into solid fats.

Visible fats: fats that can be seen, such as the fat on meat, or in the butter or oils that we use for frying or salad dressing.

Invisible fats: fats found in the products that we eat, such as biscuits, ice cream or ready meals.

Essential fatty acids: small units of fat needed to keep our bodies functioning properly.

KEY POINTS: FATS AND OILS

- Fats are used as an energy source, to protect our vital organs, to keep us warm and provide vitamins D, E and K.
- Fat gives food texture and flavour and helps fill us up.
- Saturated fats have two hydrogen atoms per carbon atom, and are mainly animal fats. They can increase cholesterol levels.
- Monounsaturated fats have one hydrogen atom to a pair of carbon atoms. They can help lower cholesterol levels.
- Polyunsaturated fats have more than one hydrogen space and will not go solid in the fridge.
- Hydrogenation adds hydrogen to oils to turn them into solid fats. These can be dangerous to our health.
- Essential fatty acids are small units of fat needed to keep our bodies functioning properly.

Exam practice questions

1. List **four** functions of fats in the body. [4 marks]

2. a) If you wanted to reduce your intake of saturated fats, which foods would you avoid eating too much of? [3 marks]

   b) What should you eat instead to provide more healthy sources of fat? [2 marks]

3. Name **three** different oils that that are mostly made up of unsaturated fats. [3 marks]

4. Name **four** foods other than fats or oils that give us a source of fat in our diet. [4 marks]

5. Explain the term 'invisible fat' and give an example of where this may be found. [3 marks]

Carbohydrates

Carbohydrates are macronutrients. They are made from carbon, hydrogen and oxygen and are mainly used to provide energy. During digestion they are broken down into a simple form called **glucose**, which can be used for energy.

There are three main forms of carbohydrates: **starch**, **sugar** and **non-starch polysaccharide (NSP)**, which is more commonly known as **fibre**.

Carbohydrates:

- provide the body with energy for physical activity
- provide the body with energy for maintaining body functions
- provide the body with fibre (NSP) to help digestion
- sweeten and flavour foods.

Carbohydrates are divided into **simple carbohydrates** and **complex carbohydrates**.

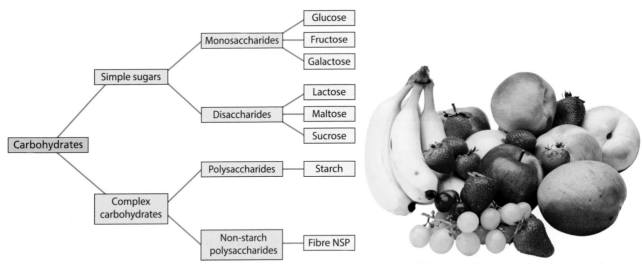

Figure 2.10 Simple and complex carbohydrates

Figure 2.11 Sources of monosaccharides in the diet

Simple carbohydrates

There are two main types of simple carbohydrates: **monosaccharides** and **disaccharides**.

Monosaccharides

These are called **simple sugars** as they are made of small molecules that are easily broken up during digestion and therefore they are absorbed quickly into the body through the wall of the digestive system, providing energy quickly.

Simple sugars include the following:

- **Glucose** is found in some fruit and vegetables. It is also available in a tablet or powdered form, which can be used by sportsmen and women to provide instant energy.

- **Fructose** is similar in structure to glucose and is found mainly in honey. It is also found in the juices of some fruit and vegetables. It is very sweet and can be used by manufacturers as a replacement for sucrose.
- **Galactose** is formed during the digestion of lactose (milk sugar).

Disaccharides

These are **double sugars** made up of two monosaccharides.

- **Sucrose** is the most common disaccharide. It is also known as cane sugar and it comes from sugar cane and sugar beet. We use the following forms of sucrose in our cooking: granulated, caster, icing and brown sugars, and also golden syrup and black treacle. It contains no nutrients and simply provides energy to the body.
- **Lactose** is found in milk.
- **Maltose** results from the fermentation of grains.

Simple carbohydrates	Examples	Foods in which they are found
Monosaccharides (simple sugars)	Glucose	Fruits and vegetables
	Fructose	Honey and some fruit and vegetable juices
	Galactose	Milk
Disaccharides (double sugars)	Sucrose	Sugar cane and sugar beet – granulated sugar, caster sugar, icing sugar, brown sugars and also golden syrup and black treacle
	Lactose	Milk
	Maltose	Grains

Table 2.2 Simple carbohydrates

Activity

Find four examples of foods containing intrinsic sugars and four containing extrinsic sugars.

The sugars we eat are in two different forms:
- **Intrinsic sugars** are found naturally in the plants we eat.
- **Extrinsic sugars** are those you can see, such as the sugar you put into cakes and biscuits or add to drinks.

Complex carbohydrates

Complex carbohydrates are called **polysaccharides**. They are made from hundreds of simple glucose molecules connected together. They provide the body with energy, but because they are so big, they take time to be broken down during digestion.

There are two main polysaccharides in our diet: **starch** and **fibre**, also known as **Non-starch polysaccharides (NSP)**.

Starch

This is found in grains, cereals, pasta and some fruit and vegetables. It fills you up for longer, so is an excellent way to stop overeating or snacking if you are trying to lose weight.

All starch comes from plant sources.

Functions of starch:

- It is broken down slowly in our digestive system to provide energy.
- It adds bulk to our diet.
- It keeps you feeling full for longer.
- Excess starch is turned into fat and stored in the body.

Fibre/NSP

This is the non-digestible part of plant cell walls that is called cellulose. It cannot be digested by our bodies, so passes straight through our digestive system, providing bulk in the diet and helping to move the waste food through the system, preventing constipation and also cleaning the walls of the digestive system to remove bacteria.

Functions of NSP:

- Holds water and keeps the faeces soft and bulky.
- Helps prevent bowel disorders including constipation, bowel cancer, diverticular disease (this is where the wall of the intestine becomes distorted and may become inflamed) and haemorrhoids (piles).
- It can help with weight control as high-fibre foods are filling, but as the fibre is not digested, it is not broken down to provide energy or calories.
- High-fibre diets have been shown to help lower blood cholesterol.

Soluble fibre slows down the digestive process and the absorption of carbohydrates, so makes us feel full for longer. It helps control blood sugar levels and can also help lower blood cholesterol levels.

Good sources are oats, beans, peas, lentils and most fruit and vegetables, particularly if you eat the skin.

Insoluble fibre absorbs water and increases bulk so keeps faeces soft, making them pass through the digestive system easily. This prevents constipation. Good sources of insoluble fibre are wholemeal bread and pasta, wholegrain cereals, brown rice and some fruit and vegetables.

Other polysaccharides

There are also three other polysaccharides in our diet:

- **Pectin** makes jams and jellies set. It cannot be digested by the body.
- **Dextrin** is formed when we toast bread or bake cakes, biscuits and pastry. We can digest this and break it down into glucose for energy.
- **Glycogen** is formed in the liver from digestion and is used as an energy source.

What happens if we eat too much or too little carbohydrate?

If you eat too much carbohydrate, the excess will be converted into fat and stored in your body. This means that you will gain weight.

The consumption of too much sugar can result in tooth decay.

If you eat too little carbohydrate, your body will use up the energy stores that it has, so you may lose weight. Your body will also use some of the protein you eat as a secondary energy source.

Activity

1 Explain why someone whose work involves heavy manual work, such as a builder, will need more carbohydrate than an office worker.

2 Why will someone running a marathon have a different type of meal to someone running a 100 metre sprint race?

Practical activity

Make a series of products using white flour and then make the same products using wholemeal flour. Taste test them on a panel and find which are preferred.

If an athlete does not eat enough carbohydrate prior to a race or a game of football, for example, they may run out of energy during the activity and begin to break down fat instead, which takes much longer. This means they will find they are not able to carry on with their activity properly, sometimes known as 'hitting the wall'.

How much carbohydrate should we eat per day?

Carbohydrates provide about 4 kcal per gram. The government Eatwell Guide advises that one third of your daily food should be starchy carbohydrates like bread, pasta, rice and potatoes. These can also provide you with fibre if wholemeal and wholegrain products are consumed.

An average 2,000 calories a day diet should contain about 250 g of carbohydrates. An equivalent for this would be a total of a bowl of wholegrain breakfast cereal, one portion of pasta and three slices of bread.

Stretch & challenge

Create a survey where you ask ten people to write down their daily food intake. Analyse each person's diet and work out if they are eating the recommended amount of carbohydrate per day.

If they are not, make suggestions to each person as to how they could increase the amount of carbohydrate they are eating.

KEY WORDS

Monosaccharides: simple sugars made of small molecules that are easily digested. They include glucose, fructose and galactose.

Disaccharides: double molecules of glucose joined together. They take longer to digest. They include sucrose, lactose and maltose.

Intrinsic sugars: sugars contained within plant cells.

Extrinsic sugars: sugars that we add to recipes, dishes we make and drinks.

Polysaccharides: complex carbohydrates made of long chains of sugar molecules that take a long time to digest. They include starch, fibre (NSP), pectin, dextrose and glycogen.

Fibre/Non-starch polysaccharides (NSP): The non-digestible part of plant cell walls. It provides bulk in the diet and helps to move the waste food through the system.

Soluble fibre: slows down the digestive process and can help lower blood cholesterol levels.

Insoluble fibre: absorbs water and helps prevents constipation.

KEY POINTS: CARBOHYDRATES

- Carbohydrates provide energy when digested.
- Monosaccharides and disaccharides are simple carbohydrates and can also be called simple sugars.
- Sugar has no nutrients and just provides energy.
- Polysaccharides are complex carbohydrates.
- Starch takes longer to break down, so fills us up for longer.
- NSP is not digested by the body, but helps with digestion and prevents constipation.
- Eating too many carbohydrates will mean we put on weight, as excess is stored as fat.
- It is recommended that one third of your daily food intake should be carbohydrates.

Exam practice questions

1 Name **two** polysaccharides. [2 marks]
2 Explain how NSP helps the digestive system to work more efficiently, and prevents constipation. [4 marks]
3 How can eating a wholegrain breakfast cereal help to prevent a student from snacking on sweet foods during the morning? [3 marks]
4 List **three** foods that are good sources of fibre. [3 marks]
5 This is a list of ingredients for an organic product:
   - Oats
   - Sugar
   - Butter (19.18 per cent)
   - Invert Sugar Syrup
   - Salt.

The nutrition label looks like this:

Typical values	Per 100 g	Per 40 g bar
Energy	1181 kJ	724 kJ
	432 kcal	173 kcal
Protein	6.1 g	2.4 g
Carbohydrates:	59.4 g	23.8 g
Of which sugars	29.9 g	11.9 g
Fat:	18.8 g	7.5 g
Of which saturated fat	9.5 g	3.8 g
Sodium	0.06 g	0.02 g
Salt	0.15 g	0.06 g

Figure 2.12 A product label

a) Identify the ingredient that provides dietary fibre. [1 mark]
b) Which **three** ingredients are carbohydrates sources? [3 marks]
c) What do you think this product is? [1 mark]

Vitamins

Vitamins are micronutrients. They are needed in small amounts by the body for a large number of different jobs.

Vitamins are referred to by letters, even though they have chemical names.

Vitamins are needed because they:
- assist in many different processes in the body, which are vital for our health
- help prevent diseases, which result from lack of certain vitamins
- are involved in the repair of cells and healing of wounds
- aid digestion and the release of energy by helping process carbohydrates
- help slow down the ageing process of cells
- help with the formation of red blood cells, healthy skin and nerves
- help with the formation of strong bones
- help with the absorption of iron.

Vitamins are divided into two main groups: **fat-soluble vitamins** and **water-soluble vitamins**.
- The fat-soluble vitamins include vitamins A, D, E and K.
- The water-soluble vitamins include the B group vitamins and vitamin C.

Fat-soluble vitamins

Fat-soluble vitamins are called this because they dissolve in fat. Vitamins A and D are added to margarine by law.

Vitamin A

Vitamin A has the chemical name retinol in animal sources and beta-carotene in plant sources.

Figure 2.13 Examples of food sources for vitamin A.

Function (what it does in the body)	Where it is found in our food	What happens if we do not eat enough (deficiency)	What happens if we eat too much (excess)
• Important for the eyesight: it supplies the retina at the back of the eye with a substance to help you see well in dim light. • Keeps the throat, digestive system and lungs moist. • It is an antioxidant: helps stop damaging substances from the air and water entering our bodies.	• Found as retinol in animal foods such as milk, cheese, butter and oily fish (tuna, herring, mackerel, sardines), liver and liver products like pâté. • Found as beta-carotene in plant foods, particularly red and orange fruit and vegetables such as carrots, apricots, red peppers, tomatoes, and also in dark green leafy vegetables such as spinach.	• We will be unable to see in dim light (called night blindness). This can eventually lead to blindness, which is a common occurrence in poor countries. • It is difficult for the body to fight disease and infections. • Children will have poor growth.	• Too much vitamin A can be poisonous. It will build up in the liver. • Pregnant women should avoid eating too much of the retinol or animal sources of vitamin A as it can cause birth defects in the developing baby.

Table 2.3 The function, sources and results of a deficiency or excess of vitamin A

Vitamin A is not destroyed by cooking processes.

Vitamin D

Vitamin D has the chemical name cholecalciferol.

Most people can get enough vitamin D from their diet and the sun. However, the following groups of people may be at risk of vitamin D deficiency:

- pregnant and breastfeeding women
- babies and young children under five, particularly if they are always covered in sun block when outside
- people who are housebound and do not get out into the sun very much
- people of African, African-Caribbean and Asian origin.

Figure 2.14 Examples of food sources for vitamin D

Figure 2.15 If young children do not get enough vitamin D they may develop rickets

Function (what it does in the body)	Where it is found in our food	What happens if we don't eat enough (deficiency)	What happens if we eat too much (excess)
<ul><li>Controls the amount of calcium that is taken from the food we eat.</li><li>Helps us to develop strong bones and teeth by assisting in the take up of calcium and phosphorus and ensuring that the bones are formed strongly during childhood and adolescence. This is because the bones have to last us the rest of our lives.</li></ul>	<ul><li>Dairy products, liver, oily fish, butter, cheese, milk and eggs.</li><li>Vitamin D is added by law to margarine.</li><li>Most of our vitamin D comes from exposure to sunlight. When the sun hits our skin a chemical reaction takes place under the skin to form vitamin D, which is then stored in the liver.</li></ul>	<ul><li>Our bodies cannot take up enough calcium and our bones and teeth will become weak.</li><li>Young children will develop a condition called rickets, where the bones bend as they are not strong enough to support the body. If this happens in older people it is called osteomalacia.</li><li>Elderly people can develop brittle bones, which break easily. This is called osteoporosis.</li><li>Vitamin D has recently been found to be important in brain function. There is a lot of research being carried out into this.</li></ul>	<ul><li>It is unlikely you will eat too much vitamin D in your diet, but if you are taking vitamin D supplements over a long period of time, it can cause more calcium to be absorbed in your body, which may then be deposited in the kidneys and damage them.</li><li>Excessive intake of vitamin D can also encourage calcium to be removed from your bones, so can weaken them.</li></ul>

Table 2.4 The function, sources and results of a deficiency or excess of vitamin D

Vitamin D is not destroyed by normal cooking processes.

Vitamin E

Vitamin E has the chemical name tocopherol.

Figure 2.16 Peanuts are an example of a food source for vitamin E

Function (what it does in the body)	Where it is found in our food	What happens if we don't eat enough (deficiency)	What happens if we eat too much (excess)
• It is an antioxidant, so helps to stop anything harmful from the air or water damaging the inside of our bodies. • It helps keep the cell walls in the body stay healthy. • It can help prevent the development of some cancers and heart disease.	Vegetable oils, lettuce, peanuts, seeds and wheat germ oil.	It is very rare to have a deficiency of vitamin E, but occasionally it can arise if a person has a problem absorbing fat, which contains vitamin E.	It is unlikely you will eat too much vitamin E from a normal diet, and there is currently no evidence to suggest that taking supplements will cause any harm to the body.

Table 2.5 The function, sources and results of a deficiency or excess of vitamin E

Vitamin E is not destroyed by normal cooking processes.

Vitamin K

Vitamin K comes from a chemical group.

Figure 2.17 Coffee: an example of a food source for vitamin K

Function (what it does in the body)	Where it is found in our food	What happen if we don't eat enough (deficiency)	What happens if we eat too much (excess)
It helps the blood to clot. This means that when we cut or injure ourselves and begin to bleed, the blood thickens and clots at the injury site to stop us losing too much blood.	• Green leafy vegetables, cheese, liver, asparagus, coffee, bacon and green tea. • It is also made by bacteria in our intestines.	• It is very rare to have a deficiency in adults. • Babies have an injection of vitamin K straight after birth.	It is unlikely you would eat too much Vitamin K in a normal diet. If you are taking vitamin K supplements, there is not enough evidence to suggest over-consumption will cause any harm.

Table 2.6 The function, sources and results of a deficiency or excess of vitamin K

Vitamin K is not destroyed by normal cooking processes.

Activity

1. Plan a meal that contains all of the fat-soluble vitamins.
2. List where all the vitamins are found.

Water-soluble vitamins

These vitamins dissolve in water. They include all of the B group vitamins and vitamin C. These vitamins cannot be stored in our bodies and so we need to eat foods which contain them every day.

B group vitamins

The B group vitamins contains several different sections. Each one has a different number, name and function in our diet.

B group vitamins	Function (what it does in our body)	Where it is found in our food	What happens if we do not eat enough (deficiency)	What happens if we eat too much (excess)
Vitamin B1 (also called thiamin)	• Helps release energy from carbohydrates. • Helps your nerves to work properly. • Helps with growth in the body.	• Cereals such as wheat and rice, cereal products and wheat germ. • Yeast and Marmite. • All types of meat. • Eggs and fish eggs (roe). • Milk and dairy food. • Seeds, nuts and beans.	• The body will have slow growth and development. • Severe deficiency causes a disease called beri-beri, which causes muscle wastage. This is often seen in poorer countries.	There is currently no evidence to suggest that eating too much will cause any harm.
Vitamin B2 (also known as riboflavin)	• Helps release energy from carbohydrates. • Helps your growth and keeps your skin healthy.	• Liver and kidneys and meats. • Eggs and milk. • Green vegetables.	• Dryness of the skin around the mouth. • Poor growth.	There is currently no evidence to suggest that eating too much will cause any harm.
Vitamin B3 (also called niacin)	• Helps release energy from carbohydrates. • Essential for healthy skin and nerves. • Can help lower levels of fat in the blood.	• Meat and poultry. • Cereals and grains. • Dairy products. • Pulse vegetables such as lentils.	Deficiency is rare, but you can develop a disease known as pellagra, which can cause diarrhoea, dermatitis (raw and rough skin) and dementia, (confusion, memory loss).	It is unlikely you will eat too much vitamin B3 in a normal diet, but taking excess supplements can cause skin flushes, and taking high doses for a long time could result in liver damage.
Vitamin B6 (also known as pyridoxine)	Helps release energy from carbohydrates.	Found in a wide range of foods.	Can lead to headaches, general aching and weakness, anaemia and skin problems.	Taking more than 200 mg a day of vitamin B6 in supplements can lead to a loss of feeling in the arms and legs known as peripheral neuropathy.
Vitamin B9 (known as folate or folic acid)	• Helps the body to use proteins. • Essential for the formation of DNA in the body cells, especially the cells which make red blood cells.	• Liver and kidneys. • Wholegrain cereals. • Pulses. • Dark green vegetables.	• Tiredness and anaemia. • Lack of folate in pregnancy can cause the foetus to have spinal malformations and spina bifida. Pregnant women normally take folic acid tablets.	Taking large doses of Folic acid can disguise a vitamin B12 deficiency, which can be a problem with older people.
Vitamin B12 (also known as cobalamin)	• Needed to form a protective coating around nerve cells to make them work properly. • Important for the production of new cells.	• Meat, fish and eggs. • Dairy products. • Vegans need to take a supplement as it is only found in animal foods.	• Tiredness and anaemia. • Muscle weakness, 'pins and needles'. • Depression and memory problems.	There is currently no evidence to suggest that eating too much will cause any harm.

Table 2.7 The function, sources and results of a deficiency or excess of the B group vitamins

All the B vitamins are destroyed by cooking in water. The loss can be reduced by steaming vegetables, or using the water the vegetables have been cooked in to make gravy or sauces to serve with the vegetables.

Vitamin C

Vitamin C is also known as ascorbic acid.

Figure 2.18 Oranges: an example of a food source for vitamin C

Function (what it does in our body)	Where it is found in our food	What happens if we do not eat enough (deficiency)	What happens if we eat too much (excess)
• Helps with absorption of iron from other foods. • Produces a substance called collagen which makes connective tissues to bind cells together in the body. • It is an antioxidant. That means it helps protect the body against polluting chemicals which can harm us.	• Fruits, especially citrus fruits such as oranges and lemons. • Blackcurrants and kiwi fruits. • Tomatoes, green leafy vegetables (not lettuce), peas, new potatoes, broccoli.	• A deficiency is rare, but if you do not eat fresh fruit and vegetables it can happen. • If you have a slight deficiency you can develop anaemia. • A severe deficiency leads to scurvy, which means you will have tiredness, bleeding gums and anaemia.	Excess vitamin C eaten in a normal food source is excreted by the body. Excess vitamin C taken as a supplement can cause nausea and diarrhoea.

Table 2.8 The function, sources and results of a deficiency or excess of vitamin C

Vitamin C is destroyed by exposure to heat and water.

Vegetables should be as fresh as possible, prepared at the last minute and cooked in as little water as possible, for as short a time as possible. Steaming is an ideal way to preserve most of the vitamin C.

Practical activity

1 Plan and cook a meal for a pregnant woman which contains all of the necessary vitamins to ensure that she is receiving all the necessary vitamins for a healthy baby.

2 Evaluate your meal and explain which foods provide which vitamins and why these are important for the health of the developing child.

KEY WORDS

Fat-soluble vitamins: vitamins that dissolve in fat. This includes vitamins A, D, E and K.

Water-soluble vitamins: vitamins that dissolve in water. This includes B group vitamins and vitamin C.

KEY POINTS: VITAMINS

- Vitamins are micronutrients that are needed for the body to function correctly, as they help to regulate chemical reactions in the body, promote health and prevent diseases.
- Vitamins are known by letters, but all have chemical names.
- A balanced diet should provide all the vitamins needed for a healthy body.
- Vitamins are either fat-soluble or water-soluble.
- Lack of any of the vitamins in the diet will mean your body cannot function correctly.

Exam practice questions

1 Identify **two** problems caused by lack of vitamin D and state what will happen to the body in each case. [4 marks]

2 Suggest **one** way of cooking green vegetables that will minimise the loss of vitamin C, and explain why this will help reduce the loss. [2 marks]

3 Name **two** sources of foods that contain folic acid, and explain why it is important that pregnant women eat enough folic acid. [4 marks]

4 How could a vegan (someone who does not eat any animal products) ensure they obtain enough vitamin B2 and B3? Design a meal to include these vitamins that could be eaten by a vegan. [6 marks]

Minerals and trace elements

Minerals and trace elements are micronutrients. They are needed for a variety of functions in the body.

Minerals include calcium, iron, magnesium, phosphorus, potassium, sodium, copper, chromium, manganese, selenium, sulphur and zinc. All of these are needed in quantities of between 1 mg and 100 mg per day.

Trace elements needed include fluoride, iodine, cobalt, molybdenum and silicon. These are needed in minute quantities, less than 1 mg per day.

- All the minerals and trace elements are necessary for the body to function properly.
- Minerals help to make strong bones and teeth, make sure we have sufficient red blood cells to transport oxygen around the body, control the amount of water in our body and make the nerves and muscles work correctly.
- Trace elements are responsible for strengthening the tooth enamel, making hormones and controlling bodily functions, and act in other muscle and nerve functions.

Mineral	Function (what it does in the body)	Where it is found in our food	What happens if we do not eat enough (deficiency)	What happens if we eat too much (excess)
Calcium	- Works with phosphorus and vitamin D to make strong, healthy bones and teeth. It is needed for normal growth in children. - Helps with blood clotting. - Involved with nerve and muscle function.	- Dairy products such as milk and cheese. - Added to white bread by law (the bread is fortified). - Oily fish, green vegetables, nuts/seeds, citrus fruits. - Soya milk, fruit juices and yoghurts may have calcium added.	- As we grow, our bones get bigger in order to bear our weight. Lack of calcium in our diets will result in poor bone structure and our peak bone mass (when our bones reach their hardest to support our bodies, during adolescence) will not be reached, causing problems of weak bones later in life. This will cause osteoporosis for elderly people. - Babies' bones will not form correctly if pregnant women do not eat enough calcium. - Blood will not clot properly after an injury.	It is unlikely you will eat too much calcium in a normal diet. Taking supplements of over 1,500 mg per day can cause stomach pain, diarrhoea and constipation.

Mineral	Function (what it does in the body)	Where it is found in our food	What happens if we do not eat enough (deficiency)	What happens if we eat too much (excess)
Iron	Needed to make haemoglobin, which is the red-coloured protein in red blood cells. Oxygen attaches itself to the haemoglobin during respiration, and it is then transported around the body to all cells to be used to provide energy. Vitamin C is needed to help with the absorption of iron in the body.	• Red meat, kidneys and liver. • Egg yolk. • Green leafy vegetables. • Dried apricots, lentils, cocoa and plain chocolate. • Some bread and cereals are fortified with iron.	• Anaemia is the most common effect of deficiency. Lack of red blood cells means lack of oxygen to the body, so tiredness and lack of energy are the symptoms. • A pale complexion, pale inner eyelids. • Weak and split fingernails.	The side effects of eating more than 20 mg of iron per day are stomach pains, nausea, vomiting and constipation.
Magnesium	Supports a healthy immune system, prevents inflammation and is involved in the digestive process.	• Green leafy vegetables. • Nuts and brown rice. • Wholegrain bread. • Fish, meat and dairy foods.	It can cause weak muscles, abnormal heart rhythm and a slight rise in blood pressure.	Too much naturally-occurring magnesium has not been shown to cause any side effects. Excess magnesium in supplements can cause diarrhoea, nausea and abdominal cramps.
Phosphorus	• Works with calcium for healthy bones and teeth. • Also involved with muscle function.	• Dairy products such as milk and cheese. • Nuts, meat and fish.	This is rare, but symptoms can be tiredness and depression.	It is unlikely you will eat too much phosphorus in a normal diet.
Potassium	Needed to help build proteins, break down carbohydrates, build muscle, control the electrical activity of the heart and maintain normal body growth	• All red meats. • Fish such as salmon and sardines. • Broccoli, peas, sweet potatoes and tomatoes. • Bananas, kiwi fruit, dried apricots. • Milk, yoghurt and nuts.	It can cause weak muscles, abnormal heart rhythm and a slight rise in blood pressure.	It is extremely unlikely you will eat too much potassium in a normal diet.
Sodium	• Maintains water balance in the body. • Involved in nerve transmissions.	• Cheese, bacon, smoked meats. • Processed foods and table salt (sodium chloride). • Monosodium glutamate and sodium bicarbonate are both additives and contain sodium.	• Unlikely as it is in so many foods, but if you are exercising in hot conditions you may get muscle cramps. • You can lose sodium if you have sickness and diarrhoea. • Too much salt can raise blood pressure and damage kidneys in babies and young children.	• Eating over the recommended amount of sodium in the diet (6 g for adults) may cause high blood pressure to develop. • Excess sodium leads to fluid retention which causes the heart to work harder. This may increase your risk of heart failure and stroke.

Mineral	Function (what it does in the body)	Where it is found in our food	What happens if we do not eat enough (deficiency)	What happens if we eat too much (excess)
Copper	Needed in minute amounts for energy production during digestion and for incorporating iron in red blood cells.	• Nuts. • Shellfish. • Offal (liver and kidneys).	Can lead to anaemia, but this is very rare.	This is extremely unlikely, but too much copper is poisonous.
Chromium	Needed in minute amounts for regulating blood sugar levels.	• Meat. • Wholegrains. • Lentils. • Broccoli. • Potatoes. • Spices.	Can cause unexplained weight loss and poor nerve function.	This is extremely rare, but can cause liver and kidney problems.
Manganese	Needed in minute amounts for bone formation.	• Tea. • Bread. • Nuts. • Cereal. • Green vegetables.	Is so rare there are only a few reported incidences.	If you take more than 400 mg a day as a supplement, you may get diarrhoea.
Selenium	Needed in minute amounts for the proper functioning of the thyroid gland.	• Brazil nuts. • Fish. • Meat. • Eggs.	This is extremely rare, but in children this can cause two diseases that affect the heart.	This is extremely rare, but can damage the nerve endings and cause numbness.
Sulphur	Needed in minute amounts for digestion and proper insulin function and creating connective tissue between the joints.	Found in most protein foods, particularly eggs.	Can result in joint pain, acne, brittle nails and hair. It may also lead to insulin resistance.	In the short term, too much sulphur will cause flatulence (wind problems) in your lower bowel.
Zinc	Needed in minute amounts to help with digestion. It also aids protein building in the body and is involved in sperm production.	• Meat. • Shellfish. • Dairy foods. • Bread. • Cereal products.	Can cause slow growth in adolescents, hair loss, diarrhoea, and can be involved in low fertility in men.	This is very unlikely, although excess consumption of supplements will reduce the amount of copper you can absorb, so may lead to anaemia.
Fluoride	Helps strengthen bones and tooth enamel.	• Sea water and fish. • It is added to our water supply.	The teeth may be more likely to develop cavities.	It is unlikely you will consume too much fluoride in a normal diet. Large doses consumed at one time in supplements can cause vomiting, nausea and abdominal pain.
Iodine	Needed to make thyroid hormones in the thyroid gland. These hormones control the metabolic rate in the body.	• Seafoods. • Milk and dairy foods, and some plants that have been grown in iodine-rich soil.	• Tiredness and lethargy. • Possible weight gain. • The thyroid gland can swell up and form a goitre.	It is unlikely you will eat too much iodine in a normal diet. Taking high doses of iodine in supplements can change the way your thyroid gland works, causing a range of symptoms including weight gain.

Mineral	Function (what it does in the body)	Where it is found in our food	What happens if we do not eat enough (deficiency)	What happens if we eat too much (excess)
Cobalt	This forms part of the structure for vitamin B12.	<ul><li>Fish.</li><li>Nuts.</li><li>Green, leafy vegetables.</li><li>Cereals.</li></ul>	Needed for support of vitamin B12, so a lack this will mean a lack of vitamin B12.	This is very unlikely, but too much cobalt over a long period could affect the heart and decrease fertility in men.
Molybdenum	This helps make and activate some enzymes in the body that repair and make genetic material.	<ul><li>Nuts.</li><li>Tinned vegetables.</li><li>Cereals.</li><li>Peas.</li><li>Green, leafy vegetables.</li><li>Cauliflower.</li></ul>	This is very rare.	This is very unlikely if you eat a normal diet, but over-use of supplements may cause joint pain.
Silicon	Keeps bones and connective tissues healthy.	<ul><li>Cereal foods.</li><li>Fruit.</li><li>Vegetables.</li></ul>	A normal diet will provide all you need.	This is very unlikely – there is not enough evidence to give any information.

Table 2.9 The function, sources and results of a deficiency or excess of minerals and trace elements

None of these minerals or trace elements are affected by normal cooking processes.

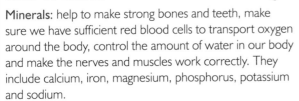

Activity

1 Select several breakfast cereal packets. Identify minerals that occur naturally in them and those that have been added to fortify the cereal.

2 Design a meal for a pregnant woman, ensuring she will have enough calcium and iron for the developing baby. Remember she must avoid liver as it is high in vitamin A, which may harm the baby's development.

KEY WORDS

Minerals: help to make strong bones and teeth, make sure we have sufficient red blood cells to transport oxygen around the body, control the amount of water in our body and make the nerves and muscles work correctly. They include calcium, iron, magnesium, phosphorus, potassium and sodium.

Trace elements: responsible for strengthening the tooth enamel, making hormones and controlling bodily functions, and act in other muscle and nerve functions. They include fluoride and iodine.

KEY POINTS: MINERALS AND TRACE ELEMENTS

- Minerals are micronutrients needed in tiny quantities.
- A balanced diet should provide all the minerals we need.
- Calcium and iron are the two main minerals we need, and they work with other vitamins and minerals.
- Trace elements are also needed to make our body function correctly.
- Normal cooking processes will not destroy minerals.

Exam practice questions

1 Identify **two** stages in our lives when it is important to eat plenty of calcium-rich foods. [2 marks]

2 Give advice to a teenager who is a vegetarian on how they can obtain enough iron in their diet. [4 marks]

3 Suggest ways you could provide sufficient calcium to a vegan who is pregnant. [4 marks]

4 Which foods should someone avoid if they are trying to reduce the amount of salt in their diet? [4 marks]

Water

Water is not a nutrient, but it is a vital part of our diet.

- Our bodies are 60 per cent water, so they need to be maintained at that level.
- Water is in all body cells and is involved in chemical reactions in the body.
- Water regulates the body's temperature, keeping it around 37°C.
- It is found in all body fluids, such as blood, urine, saliva, digestive juices and sweat.
- It helps get rid of waste products in faeces and urine.
- It keeps the linings of our digestive system, lungs, nose and throat moist.
- It helps us absorb nutrients.
- It transports nutrients, oxygen and carbon dioxide around the body.

Figure 2.19 Water is a vital part of our diet

Water is taken into the body through our drinks and in the food we eat. Fruits and vegetables have a large amount of water in them.

We should be taking in between 1.75 to 2 litres of water a day in drinks and food.

What happens if we have too little or too much water in our bodies?

If we do not drink enough, we will become dehydrated. This means you may have the following symptoms:

- headache
- dark-coloured urine
- weakness and nausea
- overheating of the body
- confusion
- changes in blood pressure.

If you drink too much water, your kidneys will not be able to cope and your blood will become diluted. Your brain will swell and this can cause nausea, convulsions and maybe death.

Stretch and challenge

Look at the number of types of bottled water available in the supermarkets. Examine the labels to see what the differences are. Work out the cost of one litre of each sort and explain why the consumer could be better off by just drinking tap water.

Fibre (Non-starch polysaccharide –NSP)

Remember that fibre is part of the carbohydrate group, but it is sometimes kept in a category by itself.

Fibre:
- helps prevent constipation by bulking out the faeces, so that they travel through the digestive system more easily
- fills you up and can help lower cholesterol levels in your blood.

Lack of fibre can cause constipation and diverticular disease and can increase the chance of cancer of the bowel.

We need a minimum of 18g of fibre per day, but the ideal amount is 30g per day.

Children need less fibre, as it will fill them up too quickly and may mean they are not getting enough of the other nutrients needed for their healthy growth.

Eating wholemeal and wholegrain products, such as plenty of fresh fruit and vegetables, particularly with the skins on, and dried fruits, nuts and seeds, will provide fibre.

Exam practice questions

1 List **three** health problems that can result from a diet low in fibre. [3 marks]

2 You are making a pasta bake for someone who is lacking fibre but does not like vegetables. How can you increase the fibre content of the dish? [4 marks]

3 How can you encourage a teenager to eat more fruit and vegetables to increase the amount of fibre and vitamins in their diet? [6 marks]

SECTION 3

Diet and good health

This section contains the following chapters:

Chapter 3 Energy requirements of individuals

Chapter 4 Plan balanced diets

Chapter 5 Calculate energy and nutritional values of recipes, meals and diets

CHAPTER 3
Energy requirements of individuals

Learning objectives

In this chapter you will learn:

- the recommended amount of energy that is provided by each of the main nutrient groups
- what percentage of each nutrient we should consume each day for a balanced diet
- how the nutrient groups work together
- how much energy you need to live, and how much you need for physical activities.

We have already looked at the different nutrient groups needed to keep your body working at its best, and how to keep yourself healthy. To recap, to ensure good health, every day we need to eat:

- protein
- fat
- carbohydrates
- vitamins
- minerals and trace elements.

Energy

Every person needs a certain amount of energy to keep the body going throughout the day. We will look at what affects the amount of energy that you need.

Energy is needed in order for our bodies to carry out every single function. Some examples of what energy is needed for are shown in Figure 3.1.

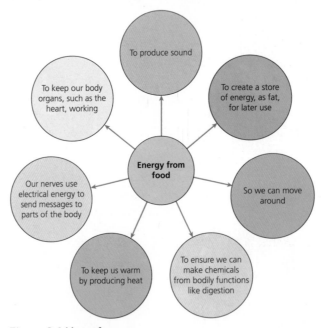

Figure 3.1 Uses of energy

How much energy is supplied from each of the nutrient groups?

- 1 gram of carbohydrate provides 15.7 Kilojoules or 3.75 kilocalories of energy.
- 1 gram of pure fat provides 37.8 Kilojoules or 9 Kilocalories of energy.
- 1 gram of pure protein provides 16.8 Kilojoules or 4 Kilocalories of energy.

The government recommendations for a healthy diet are shown in Table 3.1 below.

Nutrient	Total amount in a 2,000 kilocalorie per day diet for an adult	Percentage of energy from this nutrient
Total fat	70 grams	35%
Of which saturated fat:	20 grams	11%
Total carbohydrates	260 grams	50%
Of which sugars:	50 grams	5% from extrinsic sugars
		45% from intrinsic sugars and starches
Protein	50 grams	15%

Table 3.1 Recommended sources of energy for a healthy diet in a 2,000 kilocalorie per day diet for an adult

The **Recommended Daily Intake (RDI)** of each source of energy is discussed in Chapter 4 (see page 68). This amount will vary depending on your dietary needs, deficiencies and your lifestyle.

Vitamins and minerals are not broken down by the body, but are used by the body in the form in which they are absorbed into the body. Therefore, they are not used for energy.

The recommended amount of extrinsic sugar (remember, extrinsic sugars are those that you can see, such as the sugar you put into cakes and biscuits, or add to drinks) was reduced in July 2015 due to concerns about the obesity and dental caries problems of the population, particularly in young children. The total amount of extrinsic sugar consumed per day should be around 7 teaspoons. A regular 330 ml can of fizzy drink such as Coca Cola has about 6 teaspoons of sugar in it.

How do nutrients work together in the body?

In Chapter 2 we looked at all the main nutrient groups needed for a healthy diet and to ensure our bodies have all the things they need to work properly.

Some of these nutrients work closely together, and therefore you will get more benefit from these nutrients if you combine foods in your meals to help the body absorb these.

The ways these nutrients work together are called **complementary actions**.

The following nutrient pairs are important.

Stretch & challenge

Try to plan a meal for an adult. Now calculate what percentage of the meal is fat, how much is carbohydrate and how much is protein.

Does your meal fit in with the government guidelines above for a healthy meal?

Figure 3.2 Cereal fortified with vitamin D with milk

Vitamin D and calcium

Calcium is important for healthy bones and teeth. It is absorbed in the intestine. Vitamin D helps with this absorption, so you will gain more calcium if you eat foods containing vitamin D with them in the same meal.

An example would be a bowl of **fortified** cereal containing vitamin D and milk, which contains calcium.

Iron and vitamin C

Vitamin C helps the body to take up the **plant-based non-haem iron** that is found in vegetables such as dark green leafy cabbages and spinach. This type of iron is more difficult for the body to absorb. It is important, particularly for a vegetarian who does not eat meat, or a vegan who eats no animal products, that the amount of iron is as accessible as possible for the body.

Eating or drinking foods or drinks that contain vitamin C with a meal that has non-haem iron in it will help the body to get the amount of iron needed for a healthy body.

An example of this would be drinking a glass of freshly-squeezed orange juice with your spinach and ricotta lasagne.

Figure 3.3 Bananas are rich in potassium

Sodium and potassium

Too much sodium, or salt, in the diet will contribute to high blood pressure, stroke and heart attacks. Eating potassium helps encourage the kidneys to excrete, or get rid of, excess sodium.

Foods that are rich in potassium are sweet potatoes and bananas.

Niacin and tryptophan

Niacin is vitamin B3. Tryptophan is one of the essential amino acids that is a building block of protein, which also contains niacin. An example of this is chicken and turkey.

Vitamin B12 and folate/folic acid

These work closely together to help with cell division and replication, during foetal development and healing processes.

Zinc and copper

These are trace elements needed in very small quantities in our diet, but they compete with each other to be absorbed in the intestine. It is important to try to avoid eating foods that contain both these trace elements at the same time.

Foods containing copper include leafy greens, asparagus, wholegrains, nuts and seeds.

Foods containing zinc include seafood, meats and dairy products.

Figure 3.4 Asparagus contains copper

Dark green leafy vegetables and fats

Dark green leafy vegetables are full of trace elements and minerals. Most of these minerals are fat-soluble, so eating a healthy source of fat with greens will increase the absorption of these minerals in the body.

An example of how to do this is to toss salads in olive oil or cook greens in butter.

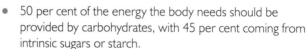

KEY WORDS

Recommended Daily Intake (RDI): the amount of each nutrient calculated to be sufficient to meet the requirements of the majority of the population of the UK. See Chapter 4, page 68 for more information.

Complementary actions: how different nutrients work with each other in the diet.

Plant-based non-haem iron: in animal foods, iron is attached to proteins called haem proteins, and it is easier to absorb. In plant foods, the iron is not attached to haem proteins, and it is harder for the body to absorb the iron.

KEY POINTS: ENERGY SUPPLIED BY THE DIFFERENT NUTRIENT GROUPS

- 50 per cent of the energy the body needs should be provided by carbohydrates, with 45 per cent coming from intrinsic sugars or starch.
- We should only be getting 35 per cent of our total energy from fats in the diet.
- Proteins should be used mainly for growth and repair, with 15 per cent of our energy needs provided from protein.
- Some nutrients work closely together in the diet to help maximise our intake of them.

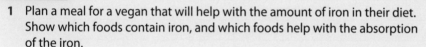

Activity

1 Plan a meal for a vegan that will help with the amount of iron in their diet. Show which foods contain iron, and which foods help with the absorption of the iron.

How much energy do we need?

The amount of energy needed just to keep everything working to stay alive is called the **Basal Metabolic Rate (BMR)**. This is simply the energy required to keep breathing, making chemicals, keep your heart beating, other body organs working, blood pumping and nerves working.

Recently the amount of energy required for your BMR has been recalculated using a scientific study organised by the government. This study also calculated the amount of extra energy needed to carry out physical activities, which are things you do like sitting, standing, walking, running, or planned and structured exercise.

These activities are called your **Physical Activity Levels (PAL)**. They will be different depending on what amount of physical activity you do.

To calculate your PAL you can use the following table.

PAL	Daily activities	Lifestyle
Less than 1.4	Hospital patient, in bed.	Inactive
1.4 to 1.55	Little physical activity at work or in leisure time.	Sedentary
1.6	Moderate physical exercise, female.	Moderately active
1.7	Moderate physical exercise, male.	Moderately active
1.7 to 2.0	Construction worker; someone who works out at the gym for an hour a day.	Moderately active
2.0 to 2.4	Physically active at work (e.g. fitness trainer).	Very active
2.4+	Professional athlete, footballer etc.	Extremely active

Table 3.2 Physical activity levels

You can increase your PAL quite easily. To increase you PAL, you will have to become more active.

Table 3.3 shows some examples of activities that increase your PAL by different amounts.

Amount of increase	Activity needed
0.15	30 minutes of moderate activity 5 or more days a week.
0.2	60 minutes brisk walking daily.
0.3	60 minutes active sport, like jogging, 5 times a week.
0.4	60 minutes jogging at 9 km per hour daily.
0.6	Intensive aerobic activity.

Table 3.3 Activities to increase PAL

Figure 3.5 You can increase your PAL by being more active

These results were published in 2011 by an organisation called the Scientific Advisory Committee on Nutrition, and are different from the previous study, which was in 1991.

The total amount of energy required for an individual person will be different to anyone else.
- Your BMR will range from 40 per cent to 75 per cent of your total energy expenditure.
- Your PAL will range from 25 per cent to 60 per cent of your total energy expenditure.

To find the amount of food you need to eat to maintain your weight (i.e. so that you do not lose or gain weight) you need to calculate your **Estimated Average Requirement (EAR)** of food, and calculate the number of calories you should consume each day.

To calculate your Estimated Average Requirement, the following calculation can be used:

Basal Metabolic Rate x Physical Activity Level = Estimated Average Requirement

BMR x PAL = EAR.

WJEC EDUQAS GCSE Food Preparation and Nutrition

The scientific study has worked out the number of calories needed per day for children on average. These are shown in Table 3.4.

Age	Basal Metabolic Rate (BMR)		Estimated Average Requirements (EAR)			
	Boys	Girls	Boys		Girls	
			Kilocalories per day			
	Kilocalories per day		Less active	More active	Less active	More active
1 year	550	500	750	790	700	750
2 years	720	660	970	1000	900	955
3 years	820	770	1100	1200	1050	1120
4 years	870	820	1250	1500	1150	1350
5 years	920	860	1330	1570	1250	1500
6 years	980	920	1400	1700	1330	1575
7 years	1030	970	1500	1760	1380	1650
8 years	1100	1030	1575	1880	1500	1750
9 years	1160	1090	1670	1980	1550	1850
10 years	1190	1100	1900	2150	1850	2050
11 years	1200	1150	2030	2250	1930	2150
12 years	1280	1200	2150	2380	2030	2250
13 years	1380	1275	2300	2570	2150	2400
14 years	1500	1330	2500	2790	2250	2780
15 years	1600	1360	2700	3000	2300	2500
18 years	1800	1400	3000	3300	2350	2600

Table 3.4 Basal Metabolic Rate (BMR) and Estimated Energy Requirement (EAR) for boys and girls aged 1 to 18 years

The scientific study also produced the figures for the EAR of men and women at different ages in their lives. These are shown in Tables 3.5 and 3.6.

Men					
Age	Height	Weight	BMR	EAR	
				Less active	More active
			Kilocalories per day		
19–24 years	178 cm	71.5 kg	1695	2530	3000
25–34 years	178 cm	71 kg	1695	2500	3000
35–44 years	176 cm	69 kg	1600	2380	2860
45–54 years	175 cm	68 kg	1600	2360	2800
55–64 years	174 cm	68 kg	1500	2360	2800
65–74 years	173 cm	67 kg	1400	2150	2500
75+ years	170 cm	65 kg	1350	2100	2500

Table 3.5 Basal Metabolic Rate (BMR) and Estimated Energy Requirement (EAR) for men

You can see from this that the older the man gets, the fewer calories he needs per day.

Activity

1 Select a children's ready meal from a supermarket. See what age range it is aimed at.

2 Now look at the number of calories that the meal contains. Work out how much of the Estimated Average Requirement (EAR) for that day is in the meal you have chosen.

3 Decide if it is a healthy meal for a child of that age.

1 Why do you think that the amount of calories needed gets less as men get older?

2 Do you think that many men eat less as they get older?

3 What happens to those men who continue to eat the same number of calories at 70 years old as they did at 30 years old?

Women				EAR	
			BMR	Less active	More active
Age	Height	Weight	Kilocalories per day		
19–24 years	163 cm	59.9 kg	1330	2000	2380
25–34 years	163 cm	59.9 kg	1330	1980	2300
35–44 years	163 cm	59.9 kg	1280	1930	2300
45–54 years	162 cm	59 kg	1280	1910	2290
55–64 years	161 cm	58 kg	1260	1880	2270
65–74 years	159 cm	57 kg	1170	1750	2050
75+ years	155 cm	54 kg	1120	1670	2000

Table 3.6 Basal Metabolic Rate (BMR) and Estimated Energy Requirement (EAR) for women

You can see that the older a woman gets, the fewer calories she needs per day.

Stretch & challenge

Ask a male teacher, your father, an older brother or a male who fits into one of the age categories in table 3.4 to write down what they ate the previous day. Then ask them what physical activities they did during that day.

Using a kilocalorie calculator on the internet, work out how many calories they ate.

1 Are they eating enough for their EAR?

2 Are they eating too many calories for their EAR?

Write up your results and give them to the person you have been studying.

KEY WORDS

Basal Metabolic Rate (BMR): the amount of energy needed to ensure the body functions are carried out before you do any physical activity.

Physical Activity Level (PAL): the amount of extra activity you do per day, including sitting, standing, running and any extra exercise you carry out during the day.

Estimated Average Requirement (EAR): the amount of calories needed per day for your body to work properly and to maintain your body weight. It is calculated by BMR x PAL.

KEY POINTS: HOW MUCH ENERGY WE NEED

- The amount of energy needed to keep everything working to stay alive is called the Basal Metabolic Rate (BMR).
- The amount of extra activity you do per day, including sitting, standing, running and exercise is called Physical Activity Level (PAL).
- The Basal Metabolic Rate (BMR) of a child will increase as they grow between birth and the age of 18 years.
- The Basal Metabolic Rate (BMR) of adults reduces slightly as they increase in age.
- To find the amount of food you need to eat to maintain your weight, you need to calculate your Estimated Average Requirement (EAR). You can use the following calculation: Basal Metabolic Rate x Physical Activity Level = Estimated Average Requirement (BMR x PAL = EAR).
- Estimated Average Requirements (EAR) of children will increase between birth and the age of 18 years.
- The Estimated Average Requirement (EAR) for girls is less than for boys.
- The Estimated Average Requirement (EAR) for adults will increase as the amount of Physical Activity Levels (PAL) increases.
- The Estimated Average Requirement (EAR) for adults decreases with age as people become less active.

Exam practice questions

1 Explain why children need more energy than an adult who works in an office and sits at a desk all day. [4 marks]

2 What activity would you suggest a 60-year-old man takes up to increase his PAL? Explain why this will help him to lose weight. [3 marks]

3 Explain why the Estimated Average Requirement (EAR) for men is higher than that for women between the ages of 19 to 24 years old. [4 marks]

4 The figures in Table 3.6 show an average height and weight measurement for women. What would happen to the Estimate Average Requirements if the women were taller than this average height? Explain why. [3 marks]

5 In Table 3.4, a more active male aged 18 years needs about 700 kilocalories more than a more active female aged 18 years. Suggest reasons why this is the case. [4 marks]

Learning objectives

In this chapter you will learn about:

■ the recommended guidelines for a healthy diet
■ how people's nutritional needs change due to their age, lifestyle choices and state of health
■ how to analyse diets
■ how to plan a balanced diet for people with specific dietary needs or nutritional deficiencies
■ how to plan balanced diets for people with specific lifestyle needs.

Recommended guidelines for a healthy diet

Most people do not have the time or the knowledge to work out exactly what they need to eat to have a balanced diet.

The government have created ways in which it is easy to see roughly how much of each food group you need to have a balanced diet.

The Eatwell Guide shows the proportion of foods you should have on your plate for every meal. The amounts are recommended for individuals over the age of 5 years.

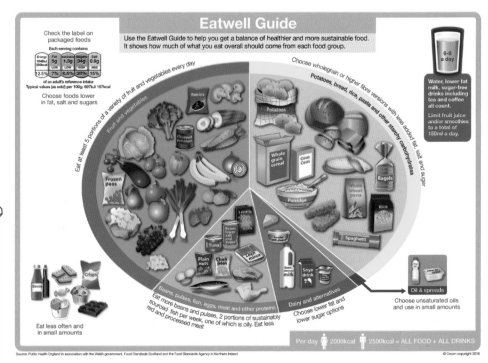

Figure 4.1 The Eatwell Guide

Activity

1 Explain why the green and yellow sections of the Eatwell Guide are the largest proportion of foods.

2 Why is the purple section the smallest?

There are also recommended amounts of each nutrient that we should be eating daily. These are called the Recommended Daily Intake (or RDI) of each nutrient, and are shown in Table 4.1.

Age	RDI (in grams)		
	Protein	Fat	Carbohydrates
1–3 years	15 g		
4–5 years	20 g		
6–10 years	28 g	70 g	220 g
11–14 years	42 g	70 g	220 g
15–18 years	55 g	70 g	230–300 g
Women	45 g	70 g	230 g
Men	55 g	95 g	300 g

Table 4.1 Recommended Daily Intake (RDI) for each nutrient

There are no Recommended Daily Intake figures of fats and carbohydrates for children under 5 years old.

As well as providing the Eatwell Guide, the government have issued a set of eight guidelines to help improve our health through our eating habits:

1 Base your meals on starchy foods.
2 Eat a minimum of five portions of fruit and vegetables every day.
3 Eat at least two portions of fish a week, one of which should be oily fish.
4 Cut down on saturated fat and sugar.
5 Eat less salt, no more than 6 grams a day for adults.
6 Drink plenty of water.
7 Do not skip breakfast.
8 Get active and try to be a healthy weight.

Figure 4.2 Base your meals on starchy foods

Basing your meal on starchy food means eating more rice, pasta, bread, potatoes, oats, millet, quinoa and cassava. These will give you vitamins, minerals and fibre, as well as providing slow-release energy (this is when the food takes a longer time to digest, so releases its energy slowly into your bloodstream). These foods will fill you up for longer, making it less likely that you will snack on sugary foods between meals.

Eating lots of fruit and vegetables provides you with essential vitamins and minerals, as well as fibre and antioxidants. One portion of fruit is about 80 grams. Fresh, frozen or canned fruit and vegetables count towards your five a day. One portion can be an apple, banana or orange. It can also be 3 heaped tablespoons of vegetables. It is easy to mix and match fruit and vegetables. A small 150 ml glass of fruit juice will also provide one portion.

Figure 4.3 A portion of all of these counts towards your recommended five a day

Figure 4.4 Eat at least two portions of fish a week, one of which should be oily fish

Two portions of fish will provide protein, vitamins and minerals for your body. White fish, such as cod, haddock and plaice is low fat. Oily fish, such as mackerel, herring, sardines, tuna and salmon will provide Omega 3 fatty acids which help to keep your heart healthy.

Cutting down on saturated fat and sugar will help prevent obesity, heart problems, dental caries (rotting teeth) and high cholesterol levels. Many foods contain hidden fats, where the fat is part of the food but you cannot see it. An example of this is in cakes and biscuits, ready meals and chocolate. These contain butter, cheese, cream, coconut oil, palm oil, meat and meat products. All of these contain saturated fat.

Figure 4.5 Cut down on saturated fat and sugar

Figure 4.6 Eat less salt – no more than 6 grams a day for adults

Cutting down on sugars will help to prevent dental problems and reduce obesity levels. Many foods contain sugar, often in a hidden form. If you look at the label of many ready meals, you will find that sugar is added for flavour and as a preservative. Sugar is present in many forms and has lots of different names: glucose, glucose syrup, dextrose, fructose, maltose, sucrose, invert sugars, corn syrup, lactose and hydrolysed starch. It is not surprising that sometimes it is difficult to see which products contain sugars!

Eating less salt will reduce your chances of developing high blood pressure and kidney problems. Salt is added to food as a preservative, and you should look out for hidden salt in foods.

A lot of foods have added salt which is not listed in the ingredients. These include bacon, smoked fish, cheese, butter, sausages and cooked meat products such as ham.

Salt is also listed on food labels under different names, e.g. sodium, sodium chloride, sodium bicarbonate, monosodium glutamate (MSG).

Drink plenty of water. Most people do not drink enough water. It hydrates the body, helps with digestion, is important for brain function and helps prevent constipation.

Figure 4.7 Drink plenty of water

This is from the food label of a ready-made spaghetti bolognese.

Look at the nutritional information and then answer the questions below.

	per 100 g of product	per pack	% of Reference Intake
Energy	548 kJ 138 kilocalories	2294 kJ 550 kilocalories	28%
Fat	6.2 grams	26 grams	37%
Saturated fat	2.8 grams	11.2 grams	60%
Carbohydrate	9.1 grams	38.2 grams	15%
Total sugars	3.1 grams	14.9 grams	17%
Starch	5.6 grams	23.3 grams	(cannot be given as a % as there is no RNI for starch, just total carbohydrates)
Fibre	6.0 grams	25.1 grams	30%
Protein	6.8 grams	28.5 grams	57%
Salt	0.53 grams	2.19 grams	37%

1   Is this product suitable for someone who is trying to cut down on the amount of saturated fat in their diet?

2   Why do you think the product has so much sugar in it? You would not add sugar to a homemade spaghetti bolognese.

3   Looking at the government guidelines for a healthy diet, would you call this product a healthy meal? Explain your reasons for your answer.

Do not skip breakfast. This means that after you have fasted, or not eaten overnight, you will break your fast. Your body needs food to give it energy after a long time without food. Research has shown that people who eat breakfast perform better at school or work, and are able to concentrate for longer.

If you think you do not have time for breakfast in the morning, there are many cereal bar products that will provide you with energy, along with a fruit smoothie or glass of milk.

Figure 4.8 Do not skip breakfast

Activity

Look at the following list of foods that could be suitable for breakfast.

- Bread, muffins, oats.
- Eggs, cheese, milk, yoghurt.
- Bacon, ham, sausages.
- Fresh fruit, dried fruit, fruit juice.
- Tomatoes, mushrooms, herbs, onions.
- Fish (for example, salmon kippers or kedgeree).

Suggest three different breakfast menus for a teenager, using these foods, that would not take very long to prepare, but would provide a filling, healthy breakfast.

Figure 4.9 Get active and try to be a healthy weight

Get active and try to stay at a healthy weight. In Chapter 3 we looked at your Physical Activity Levels, and how you should increase what you do to improve your health and fitness.

People who exercise regularly are less likely to have health problems such as obesity, heart problems and loss of bone strength as they get older.

Exercise also makes you look and feel good. It helps your digestive system to work more efficiently, develops your muscles, keeps you alert, boosts your immune system and can help to reduce your chances of developing some cancers.

It is recommended that you do a minimum of 30 minutes exercise per day. It should increase your heart rate and make you feel warm. If you do not exercise, you can become overweight, feel tired and lack energy, have weak muscles, have weak bones and develop illnesses as you get older.

Stretch and challenge

Design a healthy menu for a day and suggest an exercise routine for the following person:

- Ahmed is 40 years old.
- He works as a taxi driver. He often works late at night, so doesn't get to bed until after midnight.
- He often snacks on unhealthy food during the evening as he is sitting around waiting for people to use his taxi.
- He wakes up late and gets out of bed at 10 am. He is not often hungry, so skips breakfast.
- He does no exercise.

KEY POINTS: RECOMMENDED GUIDELINES FOR A HEALTHY DIET

- The Eatwell Guide shows the proportions of each food group that are needed to have a healthy, balanced diet.
- There are eight guidelines issued by the government to improve our health through our eating habits.

Exam practice questions

1 Identify **three** foods in the Eatwell Guide that you should eat less of to have a healthy, balanced diet. [3 marks]

2 Explain why it is important to base your diet on starchy carbohydrates. [4 marks]

3 What advice would you give someone who was trying to cut down on the amount of saturated fat in their diet? [6 marks]

Identifying how your age will change your nutritional requirements

Everyone needs to eat a balanced diet, but depending on your age, your needs will be different.

Babies and young children

From birth, babies are fed milk, either breast milk or bottle formulas. This is usually until they are between 4 to 6 months old. Breast milk is designed to give the baby everything they need, except for iron. Newborn babies are born with a store of iron, as long as the mother ate sufficient iron while she was pregnant. Milk formulas are specially designed to provide all the nutrients needed by babies.

Once the baby is old enough, he or she is introduced gradually to soft, pureed food. This is called weaning. Different foods, both cooked and raw should be introduced gradually, so the baby gets used to eating and can digest them properly.

Foods that may cause allergies, such as shellfish, peanuts, eggs and wheat products may be introduced gradually, but if there is a known allergy to these in the immediate family, it is best to wait until the baby is over a year old.

A good variety of foods should be introduced to ensure a balanced diet. Babies do not need sweet foods or added sugar, as this may create cravings later in life.

Figure 4.10 Babies are only fed breast or formula milk until between 4 and 6 months old

Children aged 1 to 4 years old

These children are usually very active, and are growing quickly. They need small, regular meals and drinks to give them energy throughout the day. The Eatwell Guide does not apply to young children. Giving them too much fibre would fill them up and mean they do not get the necessary nutrients. They also need a diet higher in fat to provide energy.

The main nutrients must be provided for the following reasons:
- Proteins, because they are growing fast.
- Fat for energy and fat-soluble vitamins. Whole milk rather than semi-skimmed should be consumed. Fat also helps with brain and nervous system development.
- Carbohydrates for energy. Try to use complex carbohydrates rather than sugar.
- Calcium and vitamin D for strong bones and teeth.
- Iron and vitamin C, for red blood cell development.
- B vitamins for digestion, nervous system development and muscle growth.

Eating habits are formed at this age, so it is important to introduce a variety of foods—not too much sugar and plenty of fruit and vegetables. Healthy eating habits learned at this age are much more likely to continue throughout their lives.
- Food can be made into fun shapes to encourage the child to try different foods.
- Children can be involved in food preparation.
- Food should be served in small portions, attractively presented.
- Eat as a family and set good examples.

Children aged 5 to 12 years

This is an age where children should be very active and are growing rapidly, so need to eat a variety of foods to provide sufficient nutrients for their bodies to be healthy.

Many children are becoming overweight and obese. This is often because they do not get enough exercise and they are eating foods that are high in fats and sugars. Allowing children to sit at computers, iPads, smartphones and televisions, and the availability of ready meals and takeaway foods, is creating a generation of unhealthy children, who will grow up to become unhealthy adults prone to many diseases.

Exercise helps the bones to become stronger and take up the calcium they need. This is laying down bone density, creating **peak bone mass**, which will help prevent problems such as weaker bones in old age.

From 5 years old, children should be following the Eatwell Guide suggestions for meals. They should be encouraged to try new foods, to extend their appreciation of food and provide all the vitamins and minerals they need.

Children should not be told to finish their food if they have eaten enough to fill them up. It is better to serve smaller portions and offer second helpings if they are still hungry, or give them a piece of fruit as an extra serving of food.

Children at this age are recommended to eat 28 grams of protein per day and around 1900 kilocalories a day for boys and 1700 kilocalories a day for girls.

Practical activity

1 Plan and make a selection of foods suitable for a child's lunch box that will provide one third of the daily recommended protein and kilocalories. The food must be attractive to encourage the child to eat it, should contain fresh fruit and vegetables and starchy carbohydrates to provide energy.
2 Invite some younger children to taste and comment on it.
3 Photograph your finished dishes and evaluate your food, using sensory descriptors.

Teenagers

This is an age where you are changing from a child into an adult. Girls will have growth spurts and will often start puberty earlier than boys. During growth spurts, both girls and boys will become hungry, so will need balanced, regular meals. Teenagers can grow several centimetres in a few months. Boys will put on large amounts of muscle tissue and therefore sufficient protein is needed to support this growth.

All the essential vitamins and minerals are needed to help the bones and internal organs form correctly. Peak bone mass is not achieved until nearly 30 years of age, so sufficient calcium and vitamin D

Figure 4.11 Teenagers need balanced, regular meals

must be eaten to ensure the risk of brittle and weak bones in older age is kept to a minimum. Milk and dairy foods will provide this, along with green leafy vegetables, cheese, nuts and seeds. Teenagers need to do weight-bearing exercise, such as aerobics, walking, dancing or gymnastics, as this can stimulate growing bones to take up minerals. Fizzy, carbonated drinks should be avoided as they may contain phosphoric acid which can stop minerals being taken up by the bones.

Teenage girls may be prone to being anaemic, as they will start their periods during this time. Iron-rich foods should be eaten, along with foods high in vitamin C to help iron absorption.

Teenagers need to avoid eating too much junk food, particularly as takeaway meals are so popular with young people. Eating breakfast will cut down on the need to snack on unhealthy foods. Providing a selection of healthy snacks, such as fruit, nuts and seeds will provide essential nutrients. Drinking water instead of fizzy, sugary drinks will keep them hydrated.

Teenagers, particularly girls, are very conscious of body image. Continual pressure from the media, peer groups and the internet can create expectations of the ideal body image, usually to be thin. This can result in under-eating, starving the body of essential nutrients and sometimes developing into conditions such as anorexia or bulimia. Anorexia is extreme weight loss and a psychological condition where the sufferer will not eat and strives to make their body as thin as possible, not recognising that they are very underweight and unwell. Bulimia is where someone will eat a lot and then make themselves sick afterwards. These are conditions that need specialist treatment.

Adults and older people

Once someone reaches adulthood they stop growing, but their body still needs all the nutrients to keep it working properly, and repair and renew cells. They also need to prevent diseases and dietary-related conditions developing.

It is important to keep the body weight within a healthy range, so a balanced diet with recommended amounts of each nutrient should be a goal for their eating habits. As a person gets older, they generally do less exercise. Their Basic Metabolic Rate will go down, so the amount of energy required will be less. They should therefore eat fewer calories to maintain their weight. This is often not easy to do, so many people find that their weight increases as they get older.

It is particularly important that older people eat sufficient calcium and vitamin D to keep their bones and skeleton healthy.

As the body gets older, the digestive and blood systems tend to slow down, and parts of the body, such as joints, start to wear out. Again, following the Eatwell Guide will help to keep the body healthy for as long as possible.

Activity

1 Think of ways you can encourage young children or teenagers to eat more fruit and vegetables.
2 What can you do to hide the fruit and vegetables, or make them more fun, tastier and more appealing to children and teenagers?
3 List the ways you would try to incorporate more of these into their diets.

The most important nutrients for older people to make sure they have plenty of are:

- Calcium and Vitamin D: The bones begin to lose minerals in older people, and a condition called **osteoporosis** can develop, where the internal structure of the bone looks as if it is full of holes. This means that the bones break easily and do not mend very well. Vitamin D must be taken, sometimes in a capsule, to ensure enough calcium is taken up.

- Fibre and water: The digestive system slows down, so constipation may become a problem. Lots of fibre-rich foods, such as wholemeal bread and pasta, fresh fruit and vegetables and wholegrain foods should be eaten. Plenty of water should be drunk to stop kidney problems developing and to avoid dehydration. Many older people are concerned about bladder problems and going to the toilet if it is difficult for them to move, however they must continue to drink plenty of water.

- Vitamin C and iron: Older people can develop **anaemia**, so lots of iron-rich foods combined with foods containing vitamin C will help to prevent this developing.

- Antioxidants: Lots of studies show that these can reduce some age-related eye conditions in older people. Antioxidants are found in vitamins A, C and E.

- Fat and sugar: Less physical activity means less energy is required, so avoidance of fatty and sugary foods will stop weight gain. Older people should try to be as active as possible to keep weight down and to help them sleep well.

- Sodium: High blood pressure often develops as we get older. Salt and sodium levels should be kept within the 6 gram per day guideline. Ready meals and processed foods often contain high sodium levels, and older people often rely on these foods as they are easy to prepare and are in convenient-sized portions. If cooking fresh meals, herbs and spices can be used for flavourings rather than salt.

- Vitamin B12: Older people often have problems with absorption of vitamin B12. This is found in milk, liver, shellfish, red meat and fortified breakfast cereals. Some studies have shown that memory loss in older people may be connected to a deficiency of the vitamin. Some older people may be recommended to take a supplement. A disease called **pernicious anaemia** may develop in older people, which means they are unable to absorb vitamin B12 through the intestine wall. In this case they will have regular injections of vitamin B12.

Other problems

Other health-related problems which may inhibit the ability to eat or swallow food are dental problems, such as where false teeth are fitted, gum and mouth problems and difficulty chewing because of Parkinson's disease or a stroke. Arthritis in the hands may mean older people struggle to cut up food or feed themselves, making eating meals difficult.

Activity

Plan a day's meals for an elderly person who has a small appetite, but likes to eat regularly and often throughout the day.

Make sure you provide all the correct nutrients needed.

KEY WORDS

Peak bone mass: the amount of bone tissue present when your skeleton has stopped growing and your bones are at their maximum strength and density.

Osteoporosis: a condition that develops when the creation of new bone does not keep up with the removal of old bone, and the bones become weak and are at risk of fracturing and breaking easily.

Anaemia: the condition that occurs when you are not making enough red blood cells to carry oxygen around your body, so you will become tired.

Pernicious anaemia: caused by low red blood cell production, which happens when the body is unable to absorb vitamin B12. Vitamin B12 is needed, with iron, to produce red blood cells.

KEY POINTS: HOW AGE CHANGES OUR NUTRITIONAL REQUIREMENTS

- Our nutritional needs change as our age changes.
- A balanced diet will provide each age group with all their nutritional needs.
- All age groups, apart from children under 5 years old, should follow the Eatwell Guide for every meal.
- Young children need to be encouraged to eat a varied diet to expand their enjoyment of food.
- Young children are growing fast and need plenty of protein for development of cells and muscles, plenty of calcium and vitamin D to achieve peak bone mass, and need to avoid unhealthy eating habits which will follow them into adulthood.
- Teenagers need balanced, regular meals. They should avoid junk food and maintain a healthy weight, get plenty of weight-bearing exercise and be encouraged to ignore peer pressure to be thin.
- Adults and older people need less energy as they age. They should keep active and eat plenty of calcium-rich foods, fresh fruit and vegetables, and drink plenty of water.

Exam practice questions

1 Why must babies have iron in their diet when they are weaned? **[2 marks]**

2 Suggest **three** ways that parents can help their young child to develop good eating habits for the future. **[3 marks]**

3 Identify **two** sources of calcium for a teenager who does not drink milk. **[3 marks]**

4 Discuss how adults can maintain a healthy weight. Suggest changes to their diet and recommend other ways that they can keep healthy. **[6 marks]**

Identifying how people's lifestyles choices and state of health change their nutritional needs

There are many choices of lifestyle that will change what we eat. There are also dietary conditions, such as allergies or nutritional deficiencies, which will have an influence on our choice of foods.

Some people develop illnesses or life-long medical conditions that mean they have to look carefully at what they eat. It is important to be aware of these conditions, and be able to plan meals that will provide a healthy range of foods that fit in with those needs.

We will look at people who make a decision to change what they eat. This could be for any of the following reasons:

- because of their religious beliefs
- because of their ethical beliefs
- because they are pregnant, or trying to get pregnant
- because they are trying to lose weight.

Different religious beliefs

There are many different religions across the world, and many of them have specific rules about what you can and cannot eat. These rules are based on beliefs, religious books or traditional celebrations.

The three main religions you will need to know about are Hinduism (Hindus), Islam (Muslims) and Judaism (Jewish).

Other religions, including Christianity, Sikhism, Buddhism and Rastafarianism, also have some dietary rules.

Table 4.2 sets out the main points of the dietary requirements for the three religions you need to know about.

Religion	Dietary requirements
Hindus (Hinduism)	<ul><li>Many Hindus are vegetarian as they do try to avoid inflicting violence and pain to any living thing. However, sometimes fish is eaten as a 'fruit of the sea'.</li><li>If they do eat meat, they do not eat beef, as cows are sacred to their religion.</li><li>They also avoid eating pork.</li><li>Eggs are forbidden in certain types of Hinduism.</li><li>Hindus believe that foods contain energies that are absorbed by people when they eat the food.</li><li>Certain foods, such as onion and garlic and alcohol, are thought to excite the body, and are forbidden. These are supposed to restrict the body from 'spiritual enlightenment', which is one of the goals of Hindu beliefs.</li><li>Hindus use a resin called asafoetida from a plant which is like celery to add flavour instead.</li><li>Cheeses made with rennet should not be eaten, but coagulated cheeses such as paneer can be eaten.</li><li>Mushrooms are also often avoided.</li><li>Dairy foods are considered to enhance your spiritual purity.</li><li>Some Hindus practise fasting.</li><li>The Hindu calendar has at least 18 feast days.</li></ul>
Muslim (Islam)	<ul><li>Muslims have a religious book called the Qur'an which sets out what is Halal, or lawful, for Muslims to eat.</li><li>This applies to the way that meat and poultry are killed before they are eaten. Halal butchers kill the animal in the following ways: a) All animals must be alive and healthy at the time of the killing. b) The animal must be killed by a Muslim with a single cut to the throat, while a special dedication or prayer is read. c) All the blood is drained from the animal after slaughter.</li><li>Unlawful, or Haram, foods include any pork or pork product, gelatine from a non-Halal-slaughtered animal, alcohol, or any foods containing alcohol, foods that contain emulsifiers made from animal fats, some types of margarine, drinks containing caffeine, and breads containing dried yeasts.</li><li>Dairy products must not contain added animal enzymes.</li><li>Ramadan is a fasting period in the ninth month of the Islamic calendar. The date will change every year, as it is based on the lunar (moon) calendar. Muslims cannot eat or drink during daylight. This is supposed to teach them self-discipline and make them think of poor people. Usually one meal is eaten at sunset and another before sunrise. Families often eat these meals together with friends.</li><li>At the end of Ramadan, the feast of Eid-ul-Fitr takes place with big celebrations. Money is donated to the poor at this time.</li></ul>
Jewish (Judaism)	<ul><li>Jews must only eat Kosher food. This means the food is clean.</li><li>Meat can be eaten from animals that chew the cud and have split or cloven hooves. This means sheep and cows.</li><li>Jews can only eat fish that have scales and fins.</li><li>Pork and shellfish is forbidden.</li><li>Animals must be slaughtered properly, allowing the blood to drain from the animal.</li><li>Dairy foods and meat must not be prepared, cooked or eaten together.</li><li>Strict Jews have different plates, cooking equipment, preparation areas and (in some cases) fridges, for dairy foods and meat.</li><li>Jews do not work on Saturdays, so food is prepared in advance.</li><li>Fasting is part of the Jewish faith. Yom Kippur, the Day of Atonement is one of the fasting days.</li><li>Other Jewish feast days include Rosh Hashanah and Passover.</li></ul>

Table 4.2 Different religious beliefs and their dietary requirements

Practical activity

One of your assessment assignments could be to produce a selection of dishes suitable for a religious festival. Select one of the religions in Table 4.2, investigate one of the feast days associated with that religion, and plan and cook a series of suitable dishes.

Evaluate your cooking, using taste tests and comments from the people who try your food.

Different ethical beliefs – types of vegetarianism

Many people change the way they eat due to their thoughts on the way animals are treated. These people usually become vegetarians or vegans.

There are several reasons why people choose a vegetarian diet:

- They do not agree with killing animals for human consumption, because they think it is a waste of resources, land, water and energy.
- They do not like the way animals are treated during the rearing and killing processes.
- They do not like the taste of meat or meat products.
- They think a vegetarian diet is a healthier option.
- They do not want to eat the hormones, chemicals and additives in meat.
- Pressure from their friends or families.
- Their religion has a vegetarian diet.

There are three main types of vegetarianism:

1 Lacto-ovo vegetarians. These people will not eat meat or fish, but they will eat animal products. This means they eat eggs, cheese and milk, and milk products such as yoghurts. Usually they want to make sure the animals have not suffered in any way, so they will eat free range eggs.

2 Lacto vegetarians. These eat the same as lacto-ovo vegetarians, but they do not eat eggs.

3 Vegans (sometimes called strict vegetarians). They do not eat meat or fish, or any animal products at all. This means no milk, cheese or eggs. They will eat only plant food. Many vegans will not wear clothes or shoes made of animal products such as leather.

Figure 4.12 Vegans do not eat any animal produce, including milk, cheese and eggs

Tip

You must not get confused with lactose intolerance, which is when people are allergic to dairy products.

Nutritional needs of vegetarians – what nutrients might be lacking in a vegetarian diet?

Iron

Iron is found mostly in meat and meat products. Vegetarians have to get their iron from vegetable sources. This is non-haem iron, which is not absorbed so easily. Vegetarians must eat plenty of vitamin C to help with the absorption of this form of iron. They must include plenty of fresh fruit and vegetables in their diet, along with fortified breakfast cereals with added iron, wholegrain cereals, dried fruits such as apricots and figs, nuts and seeds such as sunflower and pumpkin seeds, as well as lentils and beans.

Women who have heavy periods, pregnant women, someone recovering from an operation or an accident, or elderly people who are vegetarians may have to take iron supplements, so that they do not develop anaemia.

Protein

In Chapter 2 we learnt about HBV and LBV proteins. HBV proteins have all the essential amino acids needed for our bodies, but are mostly found in animal sources, apart from soya beans.

A lacto-ovo or lacto vegetarian will have no problems finding sources of HBV protein in milk, cheese and eggs. A vegan, however, will be eating mostly LBV proteins, so will have to make sure they eat a combination (or protein complementation) of these foods to get sufficient protein.

Examples of these combinations are:
- hummus with pitta bread
- nut roast made from a variety of nuts and seeds
- vegetable curry and rice
- lentil soup and wholemeal bread
- baked beans on toast

It is also possible to buy soya milk, tofu and tempeh which are made from soya beans and contain HBV protein.

Textured Vegetable Protein (TVP) also is made from soya and can be used in recipes that usually use minced beef, such as cottage pie.

Quorn™ is a trademarked product used in vegetarian foods. It is made into ready meals and sausages, burgers, fillets and escalopes. It is made from a type of fungi called mycoprotein, and egg white, so is not suitable for vegans.

Vitamin B12

This is mainly found in animal products, so again lacto-ovo and lacto vegetarians will not find it difficult to get sufficient amounts in their diets. Vegans will have to rely on fortified foods such as yeast extract (marmite), soya milk, sunflower margarine and breakfast cereals. It is possible to buy vitamin B12 supplements.

Pregnancy

During the nine months of pregnancy, a woman is providing all the nutrients to ensure the foetus develops into a healthy baby, and making sure that they stay healthy themselves.

It does not mean that they are 'eating for two'.

Activity

1 Look at the following menu. Adapt the recipe to make it suitable for a lacto-ovo vegetarian.
2 State what you are changing and why.

MENU

Starter:

Homemade tomato soup, made with chicken stock.

Main:

Lamb and vegetable kebabs with rice.

Dessert:

Lemon cheesecake, made with gelatine to set the filling.

Figure 4.13

3 Is there a way you could make this suitable for a vegan? What would you alter?

Figure 4.14 Pregnant women need some nutrients in greater quantities

All the nutrients we have discussed before are needed, but the following are needed in greater quantities:

- **Calcium**: the developing baby has to form a strong skeleton. Most of the calcium is needed in the last three months of pregnancy. The mother has to ensure she has enough calcium for herself and the baby.
- **Vitamin D** is needed to help the absorption of calcium in the body. If insufficient calcium is eaten, the baby will absorb calcium from the mother's bones, and these will become weaker.
- **Iron** forms haemoglobin, which is the part of the red blood cell that carries oxygen during respiration. During pregnancy, the volume of blood in the mother increases by up to 50 per cent. This is to help to her cope with the pregnancy, provide nutrients to the baby across the placenta in the womb, and make sure the baby is growing properly. The baby has to build up a store of iron to last for the first three months of its life, as it will get no iron from the milk it is fed.
- **Vitamin C** will help with iron absorption in the body.
- **Folic acid/folate:** this will reduce the risk of the baby developing spinal defects, such as **spina bifida**. This is a condition where the spinal nerve cord grows outside the bones of the spine. Women planning to get pregnant, and those in the first three months of pregnancy, are advised to take a folic acid supplement.
- **Fibre:** constipation may be a problem during pregnancy, as the hormones that are released in the mother's body also relax the walls of the intestine, and make the digestive system slow down. Eating bulky, fibre-rich foods will help the food to pass through and reduce constipation.

Expectant mothers should try not to put on too much extra weight. They should avoid too many fatty and sugary foods and eat a balanced, healthy diet.

Foods to avoid	Reasons why
Pâté and soft cheeses	These sometimes contain a bacteria called listeria, which can cause illness and miscarriage.
Soft blue cheeses	Can cause food poisoning.
Raw or partly-cooked eggs	Can cause salmonella poisoning.
Raw or undercooked meat	A potential risk of toxoplasmosis which is an infection caused by a parasite.
Liver; liver pâté	Liver contains a lot of vitamin A. Too much vitamin A can cause birth defects.
Certain types of fish, fresh tuna, shark, swordfish, marlin	These may contain high levels of mercury which can damage the baby's nervous system and brain.
Shellfish	Avoid raw shellfish as it can contain harmful bacteria.

Table 4.3 Foods to avoid during pregnancy

People trying to lose weight

Obesity is becoming a major health problem in the UK. Currently, about one in four adults is obese, and one in every five children aged 10 to 11 years old is obese.

The calculations used to find out if an adult is a healthy weight or obese is the **Body Mass Index (BMI)**. This means you use your weight and height measurements.

Divide your weight in kilograms by your height in metres, then divide the answer by your height in metres again.

For example: if you weigh 70 kg and you are 1.75 m tall:

divide 70 by 1.75 = 40

divide 40 by 1.75 = 22.9.

This is your BMI.

For most adults:

- A BMI under 18.5 means you are underweight.
- A healthy BMI is between 18.5 and 24.9.
- A BMI of 25 to 29.9 means you are overweight.
- A BMI of 30 to 39.9 means you are obese.
- A BMI of 40 and above means you are severely obese.

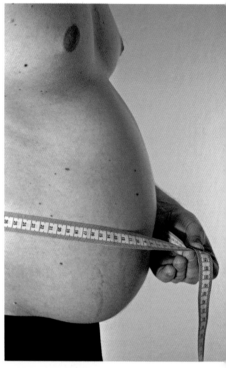

Figure 4.15 Obesity is becoming a major health problem in the UK

Being overweight can make you more likely to develop health problems such as:

- heart disease
- diabetes
- high blood pressure
- painful joints due to having to carry excess weight
- breathing difficulties
- skin rashes due to excess folds of skin
- low self-esteem and depression because of how you look.

Losing weight (if you need to) will improve your health, your self-esteem, your energy levels and improve your lifestyle.

To lose weight, you need to reduce the amount of energy in the foods you eat, and increase the amount of physical activity you do to burn up the fat stores in your body. It takes a long time to put weight on, so it will take time to lose it.

To change eating habits is often difficult. To encourage people to lose weight, they often do better in a situation where they are supported, for example, by joining a slimming club.

The following changes to eating patterns will help:

- Cut down on fatty and sugary foods that provide energy but little nutrients. Examples are cakes, biscuits, crisps, sweet fizzy drinks.
- Eat plenty of fruit and vegetables, and fibre-rich foods to fill you up, such as wholegrain cereals, wholemeal bread, brown rice and pasta.
- Grill, steam and bake food with little added fat instead of frying.
- Use lower fat versions of foods, such as cheese, milk, spreads and sauces. Check the labels on these to make sure the foods do not have added sugars.
- Use reduced sugar versions of foods.
- Increase your physical activities. Walk up stairs instead of using a lift, go swimming or to the gym.

KEY WORDS

Halal: foods that Muslims are allowed to eat.

Haram: foods that Muslims are not allowed to eat.

Kosher: foods that Jewish people can eat.

Lacto-ovo vegetarian: a person who has chosen not to eat any meat or fish, or products containing these, but will eat eggs, cheese and dairy products.

Lacto vegetarian: a person who does not eat meat, fish or eggs, or products containing these, but who will eat cheese and dairy products.

Vegan: a person who will not eat any animal product, or foods containing any animal product.

Body Mass Index (BMI): the calculation used to find out whether you are a healthy weight. A healthy BMI is between 18.5 and 24.9

KEY POINTS: HOW LIFESTYLE CHOICES AND STATE OF HEALTH CHANGE OUR NUTRITIONAL NEEDS

- Lifestyle choices change the way you eat. These can include religion, ethics, pregnancy or trying to lose weight.
- Religious rules restrict some foods in a diet by forbidding them.
- Ethical reasons often mean that vegetarianism is a life choice.
- Lacto-ovo vegetarians eat no meat or fish, but do eat eggs, cheese and milk.
- Lacto vegetarians eat no meat, fish or eggs, but do drink milk and eat milk products.
- Vegans eat no animal products at all.
- Pregnant women need to eat a diet rich in calcium, protein, iron and vitamins C and D.
- Pregnant women need to avoid certain foods to protect the growing baby from harm.
- People trying to lose weight need to cut down on fatty and sugary foods and exercise more.

Exam practice questions

1 Name **two** meats that Hindus must not eat.   [2 marks]
2 Explain what is meant by Halal food.   [3 marks]
3 Which vitamins could be lacking in a vegan diet, and which foods can vegans eat to ensure they get this vitamin in their diet?   [2 marks]

4 Discuss the foods that pregnant women should avoid, and give reasons why they should not be eaten.   [6 marks]

Planning a balanced diet for people with specific dietary needs or illnesses

In this section we will look at people who have illnesses that affect what they can eat, people with allergies that mean they have to avoid certain foods, and people with nutritional deficiencies.

- Specific dietary conditions or illnesses will restrict the foods that people can eat.
- **Coeliac disease** and **Type 2 diabetes** are conditions that require special dietary needs.
- **Cardiovascular disease** includes **coronary heart disease** and stroke.
- Anaemia, iron deficiency and **dental caries** will require changes to a diet.
- **Allergies**, such as a nut allergy or **lactose intolerance** can be life-threatening conditions, so diets need to be carefully managed.

Coeliac disease

People with this condition have an intolerance of a protein called **gluten**, which is found in wheat, barley, oats and rye. Any food product made using these ingredients will cause problems for a sufferer if it is eaten.

The lining of the small intestine is damaged by the gluten. This means that the absorption of nutrients becomes more difficult.

Symptoms of coeliac disease are:

- Weight loss, because the sufferer is not getting sufficient nutrients from the food they are eating.
- Lack of energy and tiredness, as not enough food is absorbed to provide energy.
- Diarrhoea, as the damaged intestines cannot absorb so much food, so the faeces are not made properly.
- Poor growth in children, as insufficient nutrients are absorbed.
- Anaemia can occur as not enough iron or vitamin C is absorbed.

Coeliac disease cannot be cured. It will be a life-long condition. The damage that is done to the intestine will never be properly reversed, but can be improved by eating a diet that is gluten-free.

Many products are now available in supermarkets that are gluten-free. They are clearly labelled, but are usually more expensive than other products. It is possible to have a gluten-free diet without buying special products, but it takes a lot more time and special ingredients.

Gluten-free flours will not make successful bread as the gluten is the part of wheat flour that stretches as it is kneaded. Bread that is gluten-free can be bought.

Figure 4.16 Gluten-free products

The following foods are suitable for coeliac disease sufferers:

- almond meal flour (used in bakery products as an alternative to flour)
- amaranth (used as an alternative to wheat and other cereals)
- brown, white and wild rice
- buckwheat (used to make flour and noodles)
- cassava, manioc or tapioca (used as a cereal and a thickener)
- chestnuts (ground up and used as a flour)
- coconut flour
- corn (used as a flour and thickener)
- cornstarch (used as a thickener in sauces and soups)
- guar gum (used as a thickener)
- linseeds (can be added to breakfast cereals or as an alternative to flour)
- millet (used as a muesli)
- pea flour
- polenta (used in cakes and savoury dishes)
- potato flour (used in cakes, pastries, biscuits and as a thickener)
- potatoes
- quinoa (used as an alternative to rice or couscous)
- sorghum (used as a cereal or a source of syrup)
- soy flour (used in biscuits, cakes and pastry making)
- teff (used as a flour).

Type 2 diabetes

This is called non-insulin dependent diabetes. People with this condition often develop it later in their lives, although recently, due to the sugar-rich diets of young people, more teenagers and children are being diagnosed with the condition.

The blood sugar levels in the body are controlled by a hormone called insulin, which is released from an organ called the pancreas in the body.

Continual consumption of high-value sugary foods seems to cause the pancreas to stop releasing insulin. People who are overweight or obese are also at risk of developing this disease.

A low level of glucose in the blood is called **hypoglycaemia**, and a high level is called **hyperglycaemia.**

If glucose stays in the blood it can eventually damage blood vessels in your eyes, and cause blindness.

It can restrict blood flow to your hands, feet and toes, resulting in infection and in the worst cases, amputation.

It also can cause kidney damage.

In 2014, Type 2 diabetes had been diagnosed in 3.2 million people in the UK. This number is predicted to rise to 5 million by 2025 unless diets are improved.

Type 2 diabetes can be controlled by eating a balanced, healthy diet based on complex, starchy carbohydrates.

Also the following rules should be followed:
- Cut down on sugary foods and reduce sugar in recipes. Substitute with sugar substitutes such as Canderel or Stevia.
- Use naturally sweet fruit and vegetables as a sweetener, e.g. grated carrot can be added to a cake recipe.
- Avoid dried fruits, which have concentrated sugar, and fruit juices that are concentrated, or water them down.
- Buy canned fruit in water or juice rather than syrup.
- Control the intake of fats and salt as diabetics have an increased chance of developing heart disease.
- Use lower-fat and lower-salt foods for cooking.
- Eat lean meat and poultry.
- Read food labels, looking for hidden sugars under other names.

Cardiovascular disease (CVD)

This is a general term that describes a disease of the heart or blood vessels.

Blood flow to the heart can be restricted by a build-up of fatty deposits on the walls of the arteries that supply the heart.

This causes the arteries to harden and narrow.

Coronary heart disease (CHD)

This is when the arteries supplying the heart become reduced in diameter, or blocked, and the blood cannot flow properly to the heart.

This condition is caused by high levels of **cholesterol** in the blood. Cholesterol is a fatty substance made in the liver. It is used for some functions in the body.

It attaches itself to special proteins called **lipoproteins**. It is then carried around the body in the blood.

KEY WORDS

Cardiovascular disease: any disease of the heart or blood vessels.

Coronary heart disease: when the arteries supplying the heart with oxygenated blood become partially blocked with fatty deposits from cholesterol, so the blood cannot flow freely.

If we eat a lot of saturated fats, these will form extra cholesterol which will be attached to **low density lipoproteins (LDL)**. This is sometimes called bad cholesterol. This will be deposited in the walls of the artery, building up to block the blood flow.

Unsaturated fats make less cholesterol. This attaches itself to **high density lipoproteins (HDL)**, which do not get deposited in the artery walls. This is called good cholesterol.

Other factors contribute to CHD:

- High blood pressure can put a strain on the arteries.
- Being overweight or obese puts a strain on the heart.
- Smoking cigarettes can make your blood stickier, and less likely to flow properly.

You can lower the risk of developing CHD by:

- eating more fruit, vegetables and fibre
- selecting lower-fat foods
- grilling or baking foods instead of frying
- exercising, losing weight and stopping smoking
- reducing salt intake.

Stroke

This happens when a blood vessel to the brain becomes blocked and part of the brain does not get enough oxygen. The brain cells are then damaged or destroyed. People who have a stroke can exhibit mild or severe difficulties following the event. Some may recover fully, with help and nursing care, while others may be severely impaired.

Stroke victims will need to follow the guidelines for a diet that is the same as those people with CHD.

Obesity

In the section on calculating Body Mass Index (see page 83), you can see that anyone with a BMI of over 30 is considered obese. Obese people have a much higher risk of developing health problems including:

- increased risk of coronary heart disease, high blood pressure, Type 2 diabetes, joint and mobility problems, stroke and some forms of cancer
- shortness of breath when walking or doing physical activity
- low self-esteem and depression.

Calcium deficiency

Calcium is needed for the production of healthy bones and teeth. If you lack calcium, you may develop **osteoporosis** and a calcium-deficient disease called **hypocalcaemia**.

Early-stage calcium deficiency will not cause many symptoms, but if not corrected the following symptoms can develop:

- memory loss
- muscle spasms
- numbness and tingling in the hands and feet
- depression
- hallucinations.

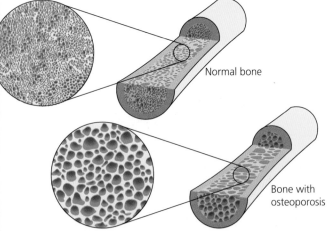
Normal bone

Bone with osteoporosis

Figure 4.17 Osteoporosis

The disease can be easily cured by eating more calcium-enriched foods, or taking a calcium supplement.

Long-term calcium deficiency will increase the risk of developing osteoporosis in older age.

Anaemia and iron deficiency

If someone does not eat sufficient iron in their diet, they will become iron deficient or anaemic. This is because they are not producing sufficient haemoglobin in their red blood cells to carry oxygen around their body.

Symptoms of anaemia include:
- tiredness and lack of energy
- shortness of breath
- a pale complexion.

Other symptoms can be:
- headache
- hair loss
- an altered sense of taste
- painful sores on the corner of your mouth
- feeling itchy
- difficulty swallowing.

People at risk of developing anaemia are:
- pregnant women
- teenage girls with heavy periods
- vegetarians and vegans.

Iron-rich foods include red meat, beans, nuts, dried fruit, wholegrains, fortified breakfast cereals, and most dark green leafy vegetables such as kale and spinach.

Remember, vitamin C is very important in helping with the absorption of iron in the intestines.

Men need 8.7 mg of iron a day. Women need 14.8 mg a day.

Dental caries

This is tooth decay, which happens when acids in the mouth dissolve the enamel on the teeth. This is a big problem in the UK, as a recent study has shown that nearly 50 per cent of 8-year-old children and 33 per cent of 5-year-old children have signs of decay in their teeth.

33 per cent of adults in the UK have some dental decay.

To help prevent dental caries:
- eat fewer biscuits, cakes and sugary breakfast cereals
- avoid giving children fizzy, sugary drinks
- water down fruit juices as these are very high in sugar

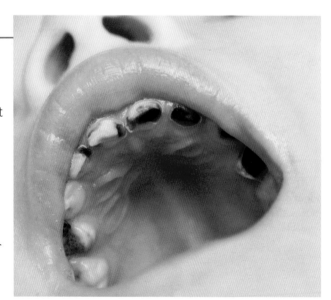

Figure 4.18 A child with dental caries

KEY WORDS

Dental caries: the rotting of teeth caused by a diet with too much sugar and insufficient cleaning of the teeth.

Allergy: when someone suffers an unpleasant, sometimes life-threatening reaction to a certain food or group of foods.

Lactose intolerance: when a person is unable to digest the sugar found in milk, which is called lactose. They cannot eat any dairy products which contain lactose.

Gluten: the protein found in wheat, barley, oats and rye products.

Figure 4.19 A large range of lacto-free products is now available

- eat healthier snacks including fruit, cheese and vegetables
- brush teeth for at least two minutes last thing at night, and again during the day
- do not brush your teeth immediately after eating, as any acid in the food you have been eating needs to be neutralised by your saliva. Wait for an hour after eating.

Allergies and intolerances

The two main food allergies and intolerances which we will deal with are nut allergies and lactose intolerance. There are also other foods that people are allergic to, including eggs, shellfish and strawberries.

Nut allergy

Allergic reactions can be minor, and often happen within a few minutes of eating the food.

- There may be a skin rash.
- The eyes can start to itch.
- The nose can become runny.
- The lips, eyelids and face might swell.
- The person may start wheezing or coughing.

An extreme reaction is when the throat starts to swell and the person cannot breathe. This is known as **anaphylactic shock**. Someone with a known allergy to nuts will often carry a special pen called an epipen, which will give them an injection of adrenaline to reduce the swelling.

Nut allergies can be life-threatening.

The presence of nuts in any environment can cause a reaction for a severe sufferer. All processed foods that are sold have to carry a nut allergy warning. Food labels have to be read carefully, and if sufferers are eating in a restaurant, they will have to check whether the food they order is nut-free.

Lactose intolerance

A person who is lactose intolerant is allergic to the sugar in milk, which is called lactose. They are unable to digest this particular sugar because they lack the correct enzyme in their small intestine. The milk sugar will pass into their large intestine, causing bloating, wind, diarrhoea and nausea. The condition is not life-threatening but is very uncomfortable for the sufferer.

People who have this condition will have to avoid dairy products.

There are now a large range of products which are lacto-free. You can buy lacto-free milk and milk products. A person with this condition can use milk alternatives such as soya, coconut, rice and almond milk.

KEY POINTS: PLANNING DIETS FOR PEOPLE WITH SPECIFIC DIETARY NEEDS OR ILLNESSES

- Identifying a specific dietary need will mean that considerations will have to be made when planning meals for people with that need.
- People with coeliac disease have an intolerance to a protein called gluten which is found in wheat, barley, oats and rye, and any products made with these ingredients. This is a life-long condition, and sufferers will have to plan a diet which does not contain gluten.
- Type 2 diabetes can develop in older age, but recently is becoming more common in younger people. The condition occurs when the body does not produce enough insulin to manage blood sugar levels.
- Cardiovascular disease is a general term for a problem with heart function and blood vessels. Eating a healthy, balanced diet will reduce the chances of developing cardiovascular disease.
- Coronary heart disease develops when fatty deposits of cholesterol partially block the arteries to the heart, causing reduced blood flow.
- High cholesterol levels can cause arteries to become blocked, causing **angina**, heart attack or stroke.
- Dental caries is caused when your diet is high in sugar, and the resulting acid created in the mouth will begin to erode the enamel coating on your teeth, causing them to rot.
- Being obese will increase your risk of developing many life-threatening conditions.
- Lack of calcium in the diet will increase your risk of developing osteoporosis as you grow older.
- Allergies will need to be managed carefully so that the foods which trigger the allergies are eliminated from the diet.
- Nut allergies can cause symptoms ranging from slight swelling of the lips and face to anaphylactic shock, when the airways swell, causing severe breathing problems.
- Lactose intolerance is when the body cannot digest the milk sugar lactose, causing bloating, wind and maybe diarrhoea.

Practical activity

1 Choose one of the two allergy conditions from above.
2 Plan a two course meal for that condition. Explain why you have chosen specific ingredients, and how they suit that person's condition.
3 Cook the meal and evaluate your finished dishes using sensory descriptors.

Exam practice questions

1 Identify **three** products from a supermarket that someone with coeliac disease should avoid. [3 marks]

2 Plan a meal for someone who has coeliac disease. Discuss the reasons why you have chosen each food ingredient for the meal and say why it is appropriate for a coeliac sufferer. [6 marks]

3 Explain what happens when someone develops Type 2 diabetes. [4 marks]

4 Give **two** foods that a person with coronary heart disease should avoid. [2 marks]

5 If someone is either deficient in iron or anaemic, suggest **three** foods they could include in their diet to improve their condition. [3 marks]

6 How could you encourage a child to eat less sugary foods to lessen their chances of developing dental caries? Suggest **four** changes to their diet. [4 marks]

7 When someone in your family has a severe nut allergy, what steps would you have to take to prevent them having a life-threatening allergic reaction? [6 marks]

Planning a balanced diet for people with high-energy needs

There are some situations when people require higher energy levels to provide sufficient calories for their lifestyle. This could be, for example:

- because of illness, and the need to build up strength
- because of a specific condition that requires a special diet
- if the person is doing a lot of exercise, e.g. a sports person or athlete.

Higher-energy diets are designed to increase the number of calories consumed per day.

Illness

If a person has had a debilitating illness, or has had major surgery and has spent a period of time when they were eating less food because of the illness, a high-energy diet will help the person to regain the weight they may have lost, and provide nutrients to aid healing and muscle growth.

Figure 4.20 All of these snack foods help to increase nutritional intake

This is particularly important in elderly people, who lose muscle tone very easily.

In these cases, small, frequent meals that are high in energy and protein are recommended. To increase the nutritional value of the meal you can, for example, add cream, butter or cheese to mashed potatoes; or add honey, jam or dried fruit to cereals. Patients could be given full-fat versions of milk and milk products instead of semi-skimmed or skimmed.

You can also provide snacks between meals to increase nutritional intake. For example:

- cheese and biscuits
- whole milk yoghurt
- milkshakes.

Elderly people with Alzheimer's disease often lose weight, so it is important to keep them on a high-energy diet to prevent weight loss. Using the same suggestions as above will help maintain weight.

Practical activity

Plan a day's menu for someone recovering from an operation in hospital. Select the lunch or evening meal you have chosen and make that dish. Evaluate your cooking using sensory descriptors.

WJEC EDUQAS GCSE Food Preparation and Nutrition

Specific dietary or genetic conditions

Some dietary or genetic conditions or illnesses mean that nutrients are not absorbed as efficiently as usual through the intestines. One of these conditions is cystic fibrosis, which is a genetically inherited condition.

Most people with cystic fibrosis do not absorb nutrients as easily as people without cystic fibrosis. They will need a diet that is higher in calories, so using full-fat versions of dairy products, increasing protein intake by having a good helping at each meal and adding oils and butter to starchy foods such as pasta and potatoes, will help to increase the total number of calories eaten each day.

People suffering from coeliac disease often do not absorb nutrients well, particularly just after they have been diagnosed with the condition, as the wall of their intestine will be damaged. They will need to increase their calorie intake, using a gluten-free diet, until their gut wall heals.

Sports people and athletes

Sports people and athletes will need to have a diet that will provide an increase in energy provision to be able to compete well and maintain their body weight.

The basic athlete's diet plan is the same as a normal person's, but the energy intake is divided into:

- 55 per cent of energy should be from carbohydrate
- 12 to 15 per cent from protein
- less than 30 per cent from fat.

Foods based on wholegrain carbohydrates should form the basis of meals, with extra carbohydrate being consumed per day depending on the level of exercise they do.

Similarly, there should be an increase in the amount of protein they eat to help post-exercise recovery and repair, as well as building extra muscle.

Figure 4.21 An extreme endurance athlete like this would need a high-energy diet to provide all the calories needed to compete and succeed in their sport

Activity	Recommended amount of carbohydrate (grams/kilogram of bodyweight)	Recommended amount of protein (grams/kilogram of bodyweight)
Light intensity exercise (30 minutes per day)	3 to 5 g/kg	1 g/kg
Moderate intensity exercise (60 minutes per day)	5 to 7 g/kg	1 to 1.2 g/kg
Endurance exercise (1 to 3 hours per day)	6 to 10 g/kg	1.2 to 1.5 g/kg
Extreme endurance (more than 4 hours per day)	8 to 12 g/kg	1.5 to 1.7 g/kg

Table 4.4 Recommended amounts of carbohydrates and protein per day for athletes

KEY POINTS: PLANNING A BALANCED DIET FOR PEOPLE WITH HIGH-ENERGY NEEDS

- There are some illnesses and genetic conditions that require high-energy diets to maintain body weight.
- Sports people and athletes require high-energy diets to provide sufficient energy to allow them to participate and succeed in their chosen sport.

Exam practice questions

1 Suggest **two** different snacks that could be offered to someone in hospital recovering from an operation, to increase their energy intake. [2 marks]

2 Explain why someone participating in a marathon race requires a higher energy intake than someone in a 100 metre sprint race. [3 marks]

3 Discuss **three** possible meals that a footballer should eat before a match, identifying the nutrients in each meal, and explaining how these would provide extra energy. [6 marks]

CHAPTER 5
Calculate energy and nutritional values of recipes, meals and diets

Learning objectives

In this chapter you will learn:

■ how to calculate the energy and main macronutrients and micronutrients in a recipe, a meal and an individual's existing diet over a period of time

■ how to use nutritional data to determine why and how to make changes to a recipe, menu and diet when necessary for a specific purpose

■ how energy balance is used to maintain a healthy body weight throughout life.

Calculating energy and nutrients

We have looked at the energy requirements of individuals in Chapter 3. We have seen that the amounts of energy needed for our Basal Metabolic Rate (BMR) alters according to our age and our Physical Activity Levels (PAL).

This affects the amount of food and energy that is needed by our bodies.

Energy is measured in kilojoules (kJ) or kilocalories (kcal).

Source of energy	Energy value in kJ	Energy value in kcal
1 g of pure carbohydrate	15.7	3.75
1 g of pure fat	37.8	9.0
1 g of pure protein	16.8	4.0

Table 5.1 Energy values of different nutrients

To calculate the energy in a particular food, multiply the number of grams of that food by the energy value in kcal for each nutrient that is in the food.

You will have to use either a computer program or a table of nutrients to find out how much of each nutrient is in that particular food.

For example: 100 g of cheddar cheese contains 25.5 g of protein, 35 g of fat and 0.1 g of carbohydrate.

Total energy from protein = 25.5 x 4 = 102 kcal

Total energy from fat = 35 x 9 = 315 kcal

Total energy from carbohydrate = 0.1 x 3.75 = 0.375 kcal

Total energy from 100 g cheddar cheese = 417.375 kcal

You can then use this figure to find out how much energy is in a recipe, by multiplying the energy amount by the number of grams you have used of that ingredient.

For example:

If you used 200 grams of cheddar cheese in a cheese sauce recipe you multiply 417.375 by 2 and the total energy from the cheese in that sauce is 834.75 kcal.

It is possible to work out individual amounts of energy for each ingredient in a recipe, a meal and a diet this way.

For example, this is how you could work out the amount of energy from a particular meal:

Lunch			
Food	Weight in grams	Calculation	Total energy provided in kilocalories
2 slices wholemeal bread	90 g 8 g protein 30 g carbohydrate 2 g of fat	8 x 4 = 32 kilocalories 30 x 3.75 = 112.5 kilocalories 2 x 9 = 18 kilocalories	162
1 slice of ham	30 g 9 g of protein 3 g of carbohydrate 5 g of fat	9 x 4 = 36 kilocalories 3 x 3.75 = 11.25 kilocalories 5 x 9 = 45 kilocalories	92.25
1 tomato	25 g 0.25 g protein 1 g carbohydrate 0 g fat	0.25 x 4 = 1 kilocalorie 1 x 3.75 = 3.75 kilocalories	4.75
1 apple	180 g 0.5 g protein 25 g carbohydrate 0 g fat	0.5 x 4 = 2 kilocalories 25 x 3.75 = 93.75 kilocalories	95.75
Carton of apple juice	150 ml 0.1 g protein 25 g carbohydrate 0 g fat	0.1 x 4 = 0.4 kilocalories 25 x 3.75 = 93.75 kilocalories	94.15
		Total energy	448.9

Table 5.2 Energy values for a meal

There are many internet dietary calculators and programs which will do this for you. Some examples include:

- **www.foodafactoflife.org.uk** (go to Resources, then Nutritional analysis for schools. Use the Explore food calculator).
- **http://www.nhs.uk/Livewell/Goodfood/Pages/Goodfoodhome.aspx** (use the calorie checker on the right-hand side)
- **www.nutracheck.co.uk**

To find the amount of macronutrients and micronutrients in an ingredient, you will also need to use special online tables, or a reference table in a book.

The websites above can be used, or your school may have printed reference tables or books which contain tables giving all the nutritional information and content of foods (for example, *The Food Bible* by Judith Williams, or *The Manual of Nutrition* published by the UK Government).

Activity

Look at the following nutrient content list for a margarita pizza.

Nutrient	Nutrient content per 100 g	Nutrient content per half of pizza
Protein	10.5 g	19.0 g
Carbohydrates	29.1 g	49.1 g
Fat	7.7 g	14.2 g

Using the energy values in kilocalories in Table 5.1, calculate the total number of kilocalories in 100 g of the pizza, and the total number of kilocalories in half a pizza.

Practical activity

Make a cottage pie using the recipe ingredients below.

Cottage pie (serves 4 people)

1 tablespoon olive oil

1 onion

500 g beef mince

400 g tin chopped tomatoes

300 ml beef stock

1 kilogram of potatoes

75 g butter

4 tablespoons of milk

Calculate the total energy in the recipe using one of the recommended websites above.

Divide your answer by 4 to find the total energy per serving.

Adapting meals and diets

We discussed and investigated many different diets in Chapter 4. People of different ages, with different lifestyles and different medical conditions or diseases, allergies and intolerances all require changes to their diets.

To keep a balanced, healthy diet it is often recommended that fat, sugar and salt is reduced and fibre is increased. There are a number of ways in which these can be done, as shown in Table 5.3.

Reduce fat	• Choose leaner cuts of meat and check for the fat content of minced beef. • Grill, bake and steam rather than fry foods. • Trim excess fat from meat. • Choose low-fat versions of spreads and dairy foods. • Reduce the amount of butter and margarine you spread on bread. • Use alternatives to high-fat mayonnaise for salad dressings. • Buy canned fish, like tuna and salmon, in brine rather than oils.
Reduce sugar	• Reduce the sugar quantity in recipes. • Use food sweeteners for stewed fruit and hot drinks. • Use alternative sweet foods such as carrots (carrot cake), ripe bananas, fresh and dried fruits to add sweetness to cakes and biscuits.
Reduce salt	• Use herbs, spices and pepper to flavour food. • Cut down on processed foods with hidden salt, like bacon, ham, cheese, salted crisps and peanuts. • Buy reduced-salt versions of foods. • Use salt alternatives, such as LoSalt, to season foods. • Read labels to identify sodium, sodium bicarbonate, monosodium glutamate and baking powder, which all contain sodium.
Increase fibre	• Eat wholegrain products, bread, cereals, pasta and rice. • Use wholemeal flour, or half wholemeal and half white, when baking. • Add oats or wheat bran to crumble toppings, pastry and other recipes. • Add dried fruit to cake recipes. • Add fresh or dried fruit to breakfast cereals. • Make smoothies with fresh fruit. • Eat the skin of fruit and vegetables, e.g. jacket potatoes. • Eat porridge for breakfast. • Add chopped vegetables to pasta sauces or cottage pie.

Table 5.3 Ways to reduce fat, sugar and salt and increase fibre

Figure 5.1 Dried fruit can be added to cake recipes or breakfast cereals to increase fibre

Practical activity

1 Choose one of the four points from Table 5.3 (i.e. reduce fat, reduce sugar, reduce salt or increase fibre).
2 Design a menu for someone using the one you have chosen.
3 Identify the ways you have altered the recipes you have chosen to make the menu fit the chosen diet. (An example might be that you are using wholemeal flour rather than white flour to increase the fibre content.)
4 Cook the meal, and try it out on members of your family. Ask them to give sensory descriptors of your food.
5 Write up your results.

If it is necessary to plan a diet for the allergies or intolerances we looked at in Chapter 4, the type of allergy will need to be identified and the appropriate foods substituted with foods that will not cause an allergic reaction.

For example, Figure 5.2 shows the normal recipe for cauliflower cheese, and an adapted recipe for someone who is lactose intolerant

Recipe 1

Cauliflower cheese: normal recipe
1 cauliflower
500 ml semi-skimmed milk
50 g margarine
50 g flour
150 g cheddar cheese

Recipe 2

Cauliflower cheese: adapted recipe
1 cauliflower
500 ml soya milk
50 g lacto-free margarine such as Pure
50 g flour
150 g lacto-free cheese

Figure 5.2 A normal recipe for cauliflower cheese, and an adapted recipe for someone who is lactose intolerant

Increasing energy intake

If you wish to increase your energy intake, for example if you are a sports person training for a specific event like a marathon, you need to increase the amount of energy you are getting from carbohydrates from 55 per cent of your total energy input to between 65 and 70 per cent.

Sports people training for an endurance or strength event, such as weightlifting, may need to increase the amount of protein they eat to build more muscle.

Energy balance

Energy balance is when we take in exactly the same amount of energy as we use every day. This will mean that someone doesn't gain or lose weight. If you eat too much, and have too much energy from the food you eat, this will be stored as fat in your body and you will gain weight.

Figure 5.3 A sports person may need to increase their energy intake

If you use more energy than you eat, you will lose weight, as you will use up energy stores from your body.

When you are young, you are probably more active than when you get older. This means that unless you eat less food as you become less active, you will gain weight.

Chapter 3 explained how to calculate energy requirements for different age groups, lifestyles and physical activities.

In order to maintain body weight **energy input must equal energy output**.

The changes in your lifestyle must be reflected in changes in what you eat, otherwise you will lose or gain weight.

KEY WORDS

Energy balance: where we take in as much energy in the food we eat as we are using up in our daily activity.

KEY POINTS: CALCULATING ENERGY AND NUTRITIONAL VALUES FOR RECIPES, MEALS AND DIETS

- Energy values, macronutrients and micronutrients can be calculated using formulas, internet calculators or reference tables in books.
- To calculate the energy in a particular food, multiply the number of grams of that food by the energy value in kcal for each nutrient that is in the food.
- To find out how much energy is in a recipe, multiply the energy amount of each food by the number of grams you have used of that ingredient. It is possible to work out individual amounts of energy for each ingredient in a recipe, meal and diet this way.
- To maintain a healthy weight, energy input must always equal energy output.

Exam practice questions

1   Look at the menu plan for a day's meals for a 30-year-old woman who is trying to lose weight. She works in an office, takes very little exercise each day, but does go to an aerobics class twice a week in the evening. Suggest changes to each meal which will reduce the kilocalories and provide a suitable alternative. State why you have made each change.

[8 marks]

2   How would you recommend that a friend, who has recently found she is lactose intolerant, can obtain sufficient calcium in her diet now that she can no longer eat dairy foods? [4 marks]

3   Identify **three** ways in which someone could reduce the amount of saturated fat in their diet. [3 marks]

4   a) State why it is important that we reduce the amount of salt we are eating. [2 marks]

    b) Explain what is meant by the term 'hidden salt'. [3 marks]

MENU

Breakfast:

Bowl of sugary cereal with semi-skimmed milk
Glass of orange juice
Cup of coffee with milk and one teaspoon of sugar

Snack:

Cup of coffee with one teaspoon of sugar
Two chocolate digestive biscuits

Lunch:

Ham and tomato sandwich on white bread
Packet of salted crisps
Can of fizzy drink

Afternoon snack:

Cup of tea with a teaspoon of sugar
Slice of fruitcake

Supper:

Beef lasagne with salad
Ice cream with strawberries
Cup of coffee with one teaspoon of sugar

Late night snack:

Cup of hot chocolate with one teaspoon of sugar

Figure 5.4 A menu plan for a 30-year-old woman who is trying to lose weight

SECTION 4

The science of food

This section contains the following chapters:

Chapter 6 The effect of cooking on food

Chapter 7 Food spoilage

CHAPTER 6
The effect of cooking on food

Learning objectives

In this chapter you will learn:

- why we cook food
- how heat is transferred through food during the cooking process
- how and why some dishes rely on more than one method of heat transference
- how to select appropriate cooking methods to conserve or modify nutritive values of foods and dishes
- why some micro-organisms, such as bacteria, are helpful in the production of our food
- what the working characteristics, functional and chemical properties of ingredients are
- reasons why dishes do not always work out correctly
- how to remedy situations when problems occur.

Why do we cook food?

No one is exactly sure how and why food started to be cooked, but the discovery of fire thousands of years ago must have been when our ancestors started cooking food.

The definition of cooking is the process of producing edible and safe foods by preparing and combining ingredients and applying heat (in most cases). Heating food causes a series of physical and chemical changes to occur. The changes may be advantageous, for example they may change the colour, texture and flavour of the food which will make it more appealing and palatable for us. However, in some cases they may be disadvantageous, for example reducing the nutritional content or generating undesirable compounds in the foods we are cooking.

Food needs to be cooked for several reasons.

I To destroy harmful bacteria in food

Many raw foods contain harmful bacteria. Eggs may contain salmonella, and meat can contain salmonella, listeria and campylobacter. Cooking to a temperature above 75°C destroys the bacteria.

Meat, fish and eggs are three foods that require cooking, although sometimes all these foods are eaten raw:

- Meat is eaten raw in steak tartare, and many people like to eat a steak that is not cooked in the middle, or 'rare'.
- Raw fish is eaten as sushi.
- Uncooked eggs are used to make homemade mayonnaise.

Figure 6.1 Raw foods can contain harmful bacteria

2 To make food easier to chew, swallow and digest

Cooking breaks down chemical bonds in proteins and carbohydrates, making meat, fish and vegetables softer and easier to chew. These processes are called **protein denaturation** and **starch degradation**.

Once the food arrives in your stomach and digestion begins, it is easier if the chemical bonds have been broken during cooking.

Protein denaturation

Many foods contain proteins: meat, fish, eggs, nuts and pulse vegetables such as lentils.

Proteins are made of amino acids, which are formed in complicated folded shapes. During heating, the bonds between the folds break and the protein starts to unravel.

This usually results in a substantial change to the food item. An example of this is what happens to an egg when it is cooked. The white of the egg turns from a clear, jelly-like substance into a rigid, white structure.

Another example is when the protein is changed to a softer structure. This can be seen when meat is cooked, for example when it is cooked in a liquid such as a stew. The tougher protein, called collagen, is broken down and reacts with water to form gelatine, which **tenderises** the meat.

Protein denaturisation is an irreversible reaction. You will never change the protein back into its original form.

Methods of cooking that result in protein denaturation are frying, boiling, grilling, steaming, baking and roasting.

To find out more about protein denaturation, see the Working characteristics, functional and chemical properties of ingredients section of this chapter (pages 128–137).

Figure 6.2 Raw egg and cooked egg showing the effects of protein denaturation

Starch degradation

Plants have cellulose cell walls to give rigidity to the plant to stop it falling over. Cellulose is a polysaccharide (we learnt about polysaccharides in Chapter 2 Macronutrients and micronutrients). When vegetables and fruit are cooked, the cell walls are broken down and the fruit and vegetables become softer. This process is called starch degradation. Think of the difference between a raw, crunchy carrot and a soft cooked carrot.

3 To develop the flavour of foods

Some foods actually taste better when they are cooked. Cooking meat develops the meaty flavour we enjoy, when it undergoes the **maillard reaction**. **Caramelisation** helps release the sweetness in starchy foods. Starch degradation also releases the sweetness in foods.

Figure 6.3 Raw and cooked carrots showing the result of starch degradation after cooking

Maillard reaction

This is a very complicated chemical reaction that occurs in foods that contain both proteins and carbohydrates, such as meat, bread, nuts, cakes and biscuits.

It is a chemical reaction between the amino acids in the protein and a sugar, such as glucose, fructose or lactose. The heat will start a range of chemical reactions, which result in the formation of a range of flavour and colour compounds.

Hundreds of different flavours are generated. For example, one of them is the savoury, meaty flavour which is characteristic of cooked meat.

The maillard reaction also results in the browning of food. The complex chemical reactions result in the production of brown-coloured compounds, which cause foods such as meat to have their characteristic brown colour when cooked.

Cooking methods that result in the maillard reaction are roasting, grilling, baking and frying.

Figure 6.4 Cooking a steak will create a savoury, meaty flavour because of the maillard reaction

Caramelisation

Caramelisation produces the desirable colours and flavours that are characteristic of food products such as coffee, confectionary, cakes, biscuits and dark beer. It changes the colour of the food to a shade of brown, which varies between light caramel to black.

This reaction occurs when foods containing a high concentration of carbohydrates are cooked at high temperatures using a dry heat, such as baking in a hot oven, roasting, grilling and frying.

i   As the food is heated, the sugars (usually sucrose), in the food melt and start to boil. The temperature at which this happens is between 110°C and 180°C.

ii Once the sucrose begins to melt it decomposes into simpler sugars, glucose and fructose.

iii A further series of chemical reactions takes place between the molecules, which results in the generation of hundreds of different flavour compounds. One of the most important is the butterscotch flavour compound, called diacetyl. Other flavours generated vary from intense sweetness to bitter, burnt flavours.

As well as changing the flavour, caramelisation causes food to brown. Table 6.1 shows how the colour becomes darker as the temperature is increased.

Caramel description	Temperature (°C)	Caramel flavour	Caramel colour
Light caramel	180	Intense, very sweet	Pale amber to golden brown
Medium caramel	180–188	Intense, sweet	Golden brown to chestnut brown
Dark caramel	188–204	Bitter, non-sweet	Very dark brown

Table 6.1 Caramelisation categories

All foods that contain carbohydrates will undergo caramelisation when heated.

Cooking methods that result in caramelisation are roasting, frying, grilling and baking.

Figure 6.5 Caramelisation causes food to brown

Starch degradation

Starch is made of large chains of glucose molecules which are joined together. When starch is heated, the bonds between the glucose molecules break, releasing the sweetness of the glucose into the food.

This reaction takes place in starchy foods such as pasta, bread, potatoes, wheat and oats.

Cooking methods that result in starch degradation are boiling, steaming, frying, grilling, baking and roasting.

4 To enable foods to rise, thicken, dissolve and set

We like to see a well-risen cake, a colourful jelly, a well-cooked quiche or a thick, glossy sauce.

- The raising agent in a cake, usually baking powder, bicarbonate of soda, or self-raising flour (which contains these two raising agents), creates carbon dioxide when it is mixed with a liquid, and the heat in the oven causes the air bubbles to expand, making the cake rise. The gluten in the flour traps the air bubbles in the mixture as it cooks.
- The gelatine in a jelly is a protein, which denatures when boiling water dissolves the jelly, and the protein sets as it cools.
- A quiche sets in the oven as the protein in the egg is denatured, and cooks as a solid.

None of these would happen if the food was not cooked. Heat causes chemical reactions to take place in our food, which allows us to enjoy the finished result.

A sauce thickens by **gelatinisation** during the cooking process.

Figure 6.6 A cake would not rise without being cooked

Starch gelatinisation

Foods containing starch are often used to thicken sauces. This is because when starch is heated in water, the starch granules absorb water and swell up. The polysaccharide called amylose leaches out of the starch granules and traps the water, causing the mixture to thicken, and form a gel. This process is called **starch gelatinisation**.

The cooking method that results in starch gelatinisation is boiling.

Figure 6.7 A roux sauce showing thickening due to starch gelatinisation

To find out more about starch gelatinisation, see the Working characteristics, functional and chemical properties of ingredients section of this chapter (pages 118–132).

5 To kill toxins and natural poisons in foods

Some foods might kill us if we ate them raw. For example, kidney beans contain toxins and need to be soaked for at least five hours, and then boiled for at least ten minutes on a high boil. Partly-cooked kidney beans can be more toxic than raw beans.

Figure 6.8 Raw kidney beans contain harmful toxins that will be destroyed during cooking

6 To make food look and smell more attractive

We all make an instant decision on whether or not we are going to eat something by how it looks on our plate. Smell also can influence our preferences. You probably would not eat a piece of raw beef, but when it is roasted you may well eat it!

Figures 6.4 and 6.5 show that caramelisation and the maillard reaction produce different colours in starchy foods and meat products when cooked, making them more attractive to us.

However, if you overcook vegetables, there can be a loss of pigmentation (colour). Foods that contain colour pigments are generally fruit and vegetables. There are four main colour pigments:

1 **Chlorophyll**, which is green. This is found in all green vegetables such as spinach, cabbage and broccoli, and fruits such as kiwi and green apples.

   Chlorophyll is a fat-soluble pigment, so may leach from vegetables if they are fried, for example in a stir-fry. Initial cooking will cause the chlorophyll to turn a brighter green, but if the food is cooked for a long time, the acids in the cells are released and the chlorophyll changes from green to a grey or olive-coloured pigment. Eventually it will turn to yellow. This can be seen if you boil green vegetables for too long.

2 **Carotenoids**, which are yellow to deep red. These are found in vegetables such as carrots, peppers and sweet potatoes, and fruit such as oranges, lemons and apricots.

   Carotenoids are also fat-soluble compounds, which means frying will cause some of the pigment to leach out into the fat. As well as this, **oxidation**, when the vegetables are exposed to the air during boiling without a lid, will cause colour loss.

3 **Anthocyanins** which are red, blue or purple. These are found in vegetables such as red potatoes, and aubergines, and fruits such as blueberries, cherries and red plums.

   Anthocyanins are water-soluble, and colours may leach out of the fruit and vegetables during boiling and soaking.

4 **Anthoxanthins** are a white pigment found in vegetables such as cauliflower. Anthoxanthins are also water-soluble. If you over boil cauliflower it turns a brown colour as the white pigment is lost.

Figure 6.9 The difference between perfectly-cooked and over-cooked vegetables.

7 To produce a variety of foods using different cooking methods

You can fry, bake or poach fish, and each way produces a different flavour to the fish. Many foods have several ways in which they can be cooked and served. Variety means we can enjoy our food.

8 To provide hot food in cold weather

It is very comforting to sit down to a plate of hot stew when the weather is very cold. Hot food also provides warmth to our bodies.

Exam practice questions

1 Describe the changes that take place when you fry an egg. Use the correct words to describe the reaction that takes place. [3 marks]

2 When a cake is cooked, the top turns brown. Explain what happens. [2 marks]

3 Identify **two** ways that green vegetables can be cooked to minimise loss of vitamin C. [2 marks]

Stretch and challenge

1 Do some investigation into the maillard reaction. Find out what temperature the reaction starts at, and exactly what happens. Why does the reaction change the flavour?

2 Do some testing on protein foods, such as chicken, to find out which temperature or cooking method produces the best colouring and taste.

3 Take photos of your results and present your findings in a short project.

KEY WORDS

Protein denaturation: where protein strands unravel and change structure to form a different shape or colour.

Starch degradation: the breaking up of starch molecules during cooking, which releases sugars and sweetness, and breaks down bonds to make the starchy food softer.

Maillard reaction: chemical reaction between proteins and carbohydrates, which changes the flavour of the food.

Caramelisation: when sugars melt at high heat to change colour to a shade of brown, and release sweetness.

Starch gelatinisation: where starch molecules swell up during boiling and absorb water to thicken a sauce. They eventually burst, creating a gel in the sauce.

KEY POINTS: WHY FOOD IS COOKED

- Food is cooked to destroy harmful bacteria, to make food easier to swallow and digest, and to develop the flavour.
- Food is cooked to enable food to rise, thicken, dissolve and set, and to kill any toxins and poisons that occur naturally in the food.
- Food is also cooked to make it look and smell more attractive, to give variety in our diet and to provide warmth to our bodies in cold weather.
- Many chemical changes take place during cooking that will alter the structure of proteins and carbohydrates, to change the appearance, colour, smell and taste of the food.

How heat is transferred to food during cooking processes

In this section we will look at how heat passes into food to cook it.

Heat is a type of energy. As heat gives energy to the **molecules** in food, they start to vibrate and move. The faster they move, the more heat is produced.

There are three ways that heat is transferred during cooking: **conduction**, **convection** and **radiation**.

Conduction

This happens when heat is directly touching a piece of equipment, or a piece of food.

If you put a metal pan on an electric or gas hob, the heat from the hob will heat up the base of the pan.

The molecules on the outside of the pan's surface will gain energy and start to vibrate. They then hit the next layer of molecules, so they start to vibrate. This process will continue through the layers of molecules, until the whole pan is hot.

When you then put a piece of meat onto the base of the pan, the molecules on the surface of the meat gain energy and start to vibrate; they hit the next layer of molecules, which start to vibrate as they gain energy. This carries on through the piece of meat until the molecules in the centre of the meat have gained energy and the meat will be hot all the way into the centre.

There are good **conductors** of heat, and bad conductors of heat.
- Metal conducts heat very well, which is why saucepans and frying pans, along with baking trays and cake tins, are made of metal.
- Water is also a good conductor of heat, which is why boiling foods works well and cooks foods quickly.
- Wood, plastic, cloth and glass are poor conductors of heat.

Convection

This only happens in liquids and gases.

The molecules of liquid or gas nearest the hot base of the pan gain heat energy, and start to rise in the pan. This allows colder liquid to move in next to the base of the pan. This colder liquid heats up and rises, allowing colder liquid to move in, and this carries on until all the liquid is heated to boiling point. As the liquid rises to the top of the pan, it will begin to cool again, so starts to drop back to the bottom, where it will be heated up again. There is a **convection current** moving in the pan. Convection currents also happen in ovens. Hot air rises and cooler air falls. A

Heat source

Figure 6.10 Conduction when cooking on a hob

convection oven uses a fan to move the heat around, so every part of the oven is approximately the same temperature.

Radiation

This occurs through space or air. Radiation transfers energy through space by invisible electro-magnetic waves. The waves are either **infra-red** or **microwaves**.

Infra-red heat waves are absorbed by the food when they reach it, and they create heat inside the food which cooks it. This happens when you put food under a grill.

Figure 6.11 Convection currents

Cooking foods in microwaves also uses radiation. The microwaves are created by a magnetron inside the oven. The microwaves are absorbed by the food, making the molecules vibrate and heat up, which then cooks the food.

Microwaves pass straight through glass, china and plastic, and do not heat them up. This is why all microwave-safe dishes are made from these materials. Metal will reflect the microwaves and damage the magnetron, so do not put metal objects into a microwave oven.

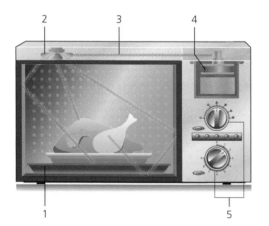

Key
1. turntable
2. metal fan to disperse the microwaves
3. microwaves being created in the magnetron and being pumped into the oven cavity
4. the magnetron
5. switches for timing and control of microwaves for different functions of the oven

Figure 6.12 Food under a grill where radiation is happening

Figure 6.13 How a microwave oven generates radiation

Explanation of how it works

The magnetron is the high voltage part of the microwave oven that produces the microwaves.

These waves are pumped into the oven cavity at a rate controlled by the timing and control switches.

The waves enter the oven cavity and are dispersed and distributed by the metal fan. They rebound from the walls of the oven, and into the food.

The turntable ensures the food is kept moving, so that it is bombarded by the waves equally all over, ensuring that it is evenly cooked or reheated.

Why some dishes rely on more than one type of heat transfer

We have already found out that there are only three ways that heat is transferred. These are conduction, convection and radiation.

If we look at the ways we cook foods, we can see how this happens.

Oven cooking/baking: The heat is transferred from the elements or gas flames into the oven by radiation. The metal tray or cake tin heats up by conduction, and the heat passes into the food that is on the tray, or in the tin, by conduction. So anything that is dry baked in an oven will rely on heat transfer this way.

Radiation → Conduction

Boiling/simmering: The heat passes from the gas ring or electric hob by conduction. This heat passes into the water by convection. The hot water heats up the surface of the food in the boiling water, for example potatoes. The heat then passes through the potato by conduction. Any food that is boiled in a liquid will use these forms of heat transfer.

Conduction → Convection → Conduction

Deep-frying/roasting/shallow-frying: The heat is transferred from the element to the oil by conduction, then through the oil by convection. The surface of the food, such as chips, or meat joints, get hot, and the heat passes through the chip, or meat joint, by conduction. Any foods cooked using these three methods will use this form of heat transfer.

Conduction → Convection → Conduction

Grilling: The heat passes through the space between the grill and the food by radiation. The surface of the food heats up, and the heat passes through the rest of the food by conduction. All foods cooked under a grill will use these forms of heat transfer.

Radiation → Conduction

Microwaving: The electro-magnetic waves move through the air by radiation. They give energy to the molecules in the food, which vibrate, causing heat which passes through the food by conduction. All fresh and ready-made foods cooked, or reheated, in a microwave use these forms of heat transfer.

Radiation → Conduction

Slow cooking: The electric element transfers heat to the pan base by conduction. The liquid in the pan heats up by convection. This makes the surfaces of the food in the slow cooker heat up, and heat passes through the pieces of food by conduction.

Conduction → Convection → Conduction

Activity

1 Think of one recipe for each method of combined heat transfer listed above.

2 Make yourself a list of recipes you have tried out and see what combination of heat transfer has happened during the cooking process for those recipes.

KEY WORDS

Conduction: this is where the heat is directly touching a food, and the heat gives energy to the molecules in the food, which start to vibrate and pass energy and heat through the food to cook it.

Convection: this happens in liquids and gases. The heat gives molecules in the liquid or gas energy, they rise and colder molecules take their place. This creates convection currents.

Radiation: happens in grilling and microwaving, when infra-red waves or microwaves pass through the air and heat the surface of the food, giving the molecules energy.

KEY POINTS: HOW HEAT IS TRANSFERRED

- Heat is a type of energy.
- As heat gives energy to the molecules in food, they start to vibrate and move. The faster they move, the more heat is produced.
- There are three ways that heat is transferred: conduction, convection and radiation.
- Conduction only happens in solids such as metal pans or food.
- Convection happens in liquids and gases.
- Radiation is when heat passes through air without the use of solid, liquid or gas.
- Many dishes rely on more than one form of heat transfer.

Exam practice questions

1 Give **three** reasons why we cook food. [3 marks]

2 Describe what energy transfers are taking place when you boil potatoes. [2 marks]

3 Explain why it is unsafe to use metal dishes in a microwave oven. [3 marks]

4 Identify **two** ways of cooking that show conduction as a method of heat transfer. [2 marks]

Selecting methods of cooking to conserve or modify nutritive values and improve palatability of foods

In this section we will look at how we can retain nutrients in our foods during cooking, how we can add extra nutrients, and how we can improve the way the food tastes and looks to make it more attractive.

Conserving nutritive value

In Chapter 2 we found out about all the nutrients that our body needs for a balanced, healthy diet, and which foods contain these nutrients.

It is not going to be any use to our bodies if we destroy some of these nutrients during the cooking process, so if you have carefully planned and selected a dish or carried out a day's menu planning to provide all the necessary nutrients, it is just as important to make sure you prepare and cook the food to supply these nutrients. This is called **conserving nutritive value**.

We also need to look at how **robust** (strong) the nutrients are before we decide on a suitable cooking method.

Nutrient	How easily is it destroyed?
Protein	Not destroyed by heat, but chemical changes result in denaturation.
Carbohydrate	Not destroyed by heat, but chemical changes result in starch degradation.
Fat	Not destroyed by heat, but some methods of cooking add fat (e.g. frying) and some reduce fat content (e.g. grilling).
Fat-soluble vitamins A, D, E and K	These will leach out of the food into the fat if foods containing these vitamins are cooked using fat.
Water-soluble vitamins B and C	These will be damaged by heat and will dissolve into the water the food is cooked in.
Minerals	Minerals are not affected by cooking processes.

Table 6.2 How easily different nutrients are destroyed

Looking at Table 6.2 we can see that the main foods that could lose nutrients during cooking are those containing vitamins. If we look back at Chapter 2, we find that many vitamins are found in fruit and vegetables, so these are the foods that we must take care of during cooking, to ensure as many vitamins are kept in the food as possible.

Table 6.3 shows which vitamins are destroyed by different cooking methods.

Nutrient	Heat	Air	Water	Fat
Vitamin A	✓			✓
Vitamin D				✓
Vitamin E	✓	✓		✓
Vitamin C	✓	✓	✓	
Thiamin (vitamin B1)	✓		✓	
Riboflavin (vitamin B2)			✓	
Vitamin B6	✓	✓	✓	
Folate	✓	✓		
Vitamin B12	✓		✓	

Table 6.3 Which vitamins are lost by which cooking methods

Fat-soluble vitamins

Vitamin A is found in animal products such as liver, milk, cheese and oily fish as retinol, and in fruit and vegetables such as carrots, red peppers and dark green, leafy vegetables as beta-carotene. If we fry these foods, we will lose the vitamin A, as it will dissolve into the fat we are frying in.

This will not be so much of a problem if you are serving the fat as part of the dish, for example a stir-fry uses a minimum amount of fat, and the food is coated with the fat as it is cooked, so is usually eaten as part of the dish.

However, if we are frying food, such as liver or oily fish, the liver or fish is removed from the fat before we serve it. This means some of the vitamin A will be left in the frying pan.

Vitamin D is found in oily fish, butter, cheese and milk. It is also added by law to margarine.

Vitamin E is found in vegetable oils, peanuts, seeds and wheat germ oil.

Vitamin K is found in cheese, liver, asparagus and leafy vegetables.

It is unlikely that cooking these foods will result in loss of these vitamins, as they are needed in small quantities, and we usually eat enough foods to give us sufficient quantities of these vitamins.

Water-soluble vitamins

These are much more likely to be destroyed by cooking methods as they are not very robust. They are easily destroyed by heat, and dissolve in water during cooking.

Vitamin C is also destroyed when it is exposed to oxygen, so when you shred spinach or cabbage, the vitamin C starts to break down as oxygen from the air hits it.

Figure 6.14 This stir-fry will serve the fat it is cooked in as part of the meal.

Activity

Can you think of an alternative way to cook liver or oily fish that does not involve frying in fat?

Figure 6.15 The vitamin C in these vegetables is easily destroyed by heat and will dissolve in water

Ways to ensure as much of these vitamins remain in the food as possible:

- Buy the foods as fresh as possible, as these will have the most vitamin C.
- Prepare them at the last possible minute to reduce the amount of time they are exposed to the air.
- Select a cooking method that uses as little water as possible; steaming rather than boiling, microwaving, or frying in a little oil.
- Cook them for as short a time as possible and serve immediately as the longer you keep them hot, the more vitamin C will be destroyed.
- If you have boiled the vegetables, you can use the cooking liquid to make a sauce or gravy to serve with the vegetables, or make a soup from the liquid.
- Root vegetables such as potatoes or sweet potatoes will retain about 65 per cent of their vitamin C if they are baked whole.

Practical activity

1   Choose a selection of green, leafy vegetables, for example spinach, kale and cabbage.

Shred the leaves and cook them in the following ways:

a) Boil some in a deep pan of water.
b) Boil some in a small amount of water.
c) Steam some in a steamer.
d) Microwave some.
e) Stir-fry some in a wok.

Compare the results, looking at texture, colour and taste.

Which do you prefer?

Is there any difference in taste and texture?

The steamed, microwaved and stir-fried vegetables should contain more vitamin C than the other two boiled vegetables.

Modifying nutritive value

When a recipe or meal plan is selected, it is possible to change the amount of any nutrient by adding or removing ingredients.

In Table 5.2 in the previous chapter we looked at ways to reduce salt, sugar, fat or increase fibre in the diet. Similarly, extra nutrients can be added to food to cater for any needs.

For example:

- adding an egg to a product such as creamed potatoes to enrich it and add extra protein
- adding grated vegetables, such as carrots, into a meat dish like lasagne, to give vitamins and fibre
- adding chopped meat, such as bacon, to a macaroni cheese for extra protein
- adding powdered products, such as wheat germ to stews and soup for extra protein.

Improving palatability of a dish

The taste buds on our tongues recognise, salt, sour, sweet, bitter and umami (a pleasant savoury taste).

There are times when food **palatability** needs to be improved.

- When elderly people have small appetites, it is important to make the food as tasty as possible, to encourage them to eat. This will also have the benefit of increasing saliva flow in the mouth, so the food is digested more easily, and mouth problems are reduced.

- When young or elderly people have difficulty chewing, it is possible to tenderise meats and make them easier to chew, by beginning the denaturation process of the protein in the meat by hitting it first with a meat mallet. This starts to break the bonds between the proteins, and makes it tenderer when cooked, so it is easier to chew.

- People undergoing some cancer treatments, such as chemotherapy, may experience problems with their mouths, and food may not taste very nice. As it is very important that these people maintain a healthy diet, food additives such as slight sweeteners can be put in to help improve the taste of food, making it more palatable to the person.

- Adding fat to foods can increase palatability, as most people enjoy the **mouthfeel** (the creamy satisfying feel) of fatty foods, so tend to eat more and enjoy them more. Many manufacturers add fat to foods to improve the flavour.

Figure 6.16 Meat being hit by a meat mallet

KEY WORDS

Conserving nutritive value: how to stop nutrients being lost from foods during the cooking process.

Palatability: the way the food tastes and its appeal to us.

Mouthfeel: a term used to explain the creamy, flavourful taste and sensation created by fats in a recipe.

KEY POINTS: CONSERVING OR MODIFYING NUTRITIVE VALUES AND IMPROVING PALATABILITY OF FOODS

- Nutrients can be lost during the cooking process.
- The most likely nutrients to be lost are water-soluble vitamins.
- Fat-soluble vitamins are vitamins A, D, E and K.
- Water-soluble vitamins are vitamins B and C.
- Steaming and microwaving will reduce loss of water-soluble vitamins.
- In certain situations it is necessary to improve the palatability of a dish to ensure sufficient amounts of nutrients are eaten.

Exam practice questions

1   Identify **two** ways that Brussels sprouts can be cooked to preserve the vitamin C content.   [2 marks]
2   Explain why it is important to make food for elderly people as palatable as possible.   [4 marks]
3   How would you alter a recipe to add more protein for someone who was recovering from an operation and needed to have extra protein for healing? State the recipe that you would use and what you would add to it. Describe how this would increase the protein content.   [4 marks]

The positive use of micro-organisms

Most people think of micro-organisms as harmful, and indeed they can be. A lot of time is spent making sure that cross-contamination of harmful bacteria and moulds does not happen during the preparation and cooking processes.

However, some micro-organisms are useful during food production. We would not have many of our favourite foods unless they were used.

Dairy products

Most people love cheese, and many of us eat yoghurt. Both of these products are made from milk. In order to turn milk into cheese and yoghurt, a bacteria needs to be added.

Making yoghurt:

● Bacteria feed on the lactose (sugar) in the milk, and convert it into lactic acid, which denatures the milk proteins and thickens the product.
● The acid gives yoghurt its sharp taste.
● Flavours and colours are often added to the yoghurt before it is packaged.

Making cheese:

● The starting process for making cheese is the same as for yoghurt.
● After this, an enzyme called **rennet** is added, which separates the milk into curds and whey. The curd, which is the thick, white part, is used to form the cheese.
● Different flavours and processes make all the different cheeses that we like to eat.
● Blue cheeses, such as stilton or Roquefort cheese, have moulds added to them to create the veins of blue that are found in the cheeses. The moulds are deliberately introduced to the cheese by piercing the cheese with metal devices (copper wire) to allow air to enter the holes and encourage the blue mould to grow.

Figure 6.17 Yoghurt uses micro-organisms during its manufacturing process

Figure 6.18 Stilton cheese has blue mould growth

Practical activity

1 Carry out some research to find a recipe for making yoghurt.
2 Plan, prepare and make a batch of yoghurt.
3 Split the yogurt into three pots and add three different flavours.
4 Ask other members of your group to taste test the samples; use sensory descriptors to help you identify the strengths and areas to improve.

Meat products

Salami, chorizo and French saucisson are made from **fermented** meat. Fermentation is the chemical breakdown of a substance by micro-organisms.

In these products, bacteria are used to change the acidity of the meat and prevent harmful bacteria spoiling it. This results in denaturation of the protein. Salt or flavourings are added, and then the products are left to dry in the air at controlled low temperatures.

Figure 6.19 Salami and chorizo are produced using helpful bacteria

Fermentation of sugar

Alcoholic drinks use yeast and the resulting sugar fermentation to produce the alcohol.

For hundreds of years, people around the world have made forms of naturally-fermented sweet, fizzy drinks, such as ginger beer, dandelion and burdock, elderflower fizz, and other flavoured soda drinks.

If you make your own fizzy drinks at home, yeast and sugar is used to create a 'starter', which will encourage lactic acid bacteria to grow. This will eventually create the carbon dioxide, which makes the drink fizzy.

The fizzy drinks that we consume today have carbonated, fizzy water added to the base syrup before the drink is canned or bottled.

Practical activity

Find a recipe for a homemade fizzy drink, and see if you can create your own version of your favourite canned drink.

Bread making

Yeast is the product used in bread to make it rise. Yeast is a micro-organism, but one that is not harmful to us. Yeast is also used in beer making and wine making.

Quorn™

Quorn™ is a meat substitute that can be eaten by ovo-lacto vegetarians and lacto vegetarians, but not vegans. Quorn™ is made from a mycoprotein, which is derived from a fungus. The protein is mixed with egg white and formed into shapes such as sausages, burgers, meat-sized chunks and Quorn™ mince.

> **KEY POINTS: POSITIVE USE OF MICRO-ORGANISMS**
> - There are many micro-organisms that are beneficial to us.
> - The foods produced using micro-organisms include cheese, yoghurt and meat products such as salami and chorizo.
> - Yeast is a micro-organism used in bread, fizzy drinks, wine and beer production.
> - Quorn™, a meat substitute, is made from a micro-organism.

Working characteristics, functional and chemical properties of ingredients

In this section we will find out:

- why we select certain ingredients for recipes, and how their use makes the product turn out in a successful way
- the **functional properties** of the ingredients we use (this means that there is a particular reason that we use an ingredient in a recipe). It could be for its taste, texture or colour for example
- the chemical reactions that take place during the cooking of certain ingredients, for example, baking it makes it set.

The three macronutrients – protein, carbohydrates and fats – make up the basis of nearly all our recipes and meals. Each of these has particular characteristics, which mean that when they are cooked, they alter.

To recap, when food is cooked:

- it changes colour
- it changes texture
- the flavour of the food changes.

We will look more closely at what different properties of these three macronutrients make them suitable for certain recipes, so we can have a successful result.

Proteins

Proteins exhibit the following characteristics:

- **coagulation**
- **foam formation (or aeration)**
- **gluten formation**
- **denaturation.**

We discovered that when proteins are heated, the bonds between the amino acids unravel and create a different structure. This is called **denaturation**. (Remember, this is when the bonds between the folds in the amino acids break and the protein starts to unravel, forming different shapes, see Figure 6.2.)

This will happen when the protein:

- is heated
- is whisked
- is put in an acid, for example, vinegar.

The main protein source that we use in cooking to create different recipes is eggs. Eggs have many properties and can be both cooked and raw in recipes. They demonstrate coagulation, foam formation and denaturation.

Table 6.4 shows how and where eggs exhibit these properties.

The function of eggs	Why the egg is used	Examples of recipes using this function
To bind ingredients together	The protein in the egg will coagulate (become thick and set) when heated, so it holds the other ingredients together.	Fish cakes Potato croquettes
To trap air	The protein in the egg will stretch when it is whisked (denaturation) and this will allow air to be trapped as little bubbles (foam formation or aeration). When the product is cooked it will solidify around the air.	Meringues Whisked sponges, such as Swiss rolls Mousses and soufflés
To thicken products	As the protein in the egg is heated it coagulates and thickens the product.	Egg custard Quiche Lorraine
To coat products	The product is dipped in egg and breadcrumbs before it is fried. The egg coagulates and seals the product in the crispy coating.	Scotch eggs Fish in breadcrumbs
To create an emulsion	The protein in the egg will stop oil and water separating. This is called an emulsion.	Mayonnaise, when the oil and vinegar are held together by the egg yolk
To glaze products	Egg is brushed on the surface of a product, and when it is cooked the egg will form a shiny surface (coagulation).	Pastry toppings on pies Bread rolls or a loaf of bread

Table 6.4 Properties of eggs in cooking

Eggs exhibit the following properties that are unique to proteins:

- They can coagulate (set) when heated.
- They can trap air by forming a foam, which will set when cooked.
- The protein denatures (unravels and forms new structures) when heated.

Practical activity

One of your assessment tasks could be to make a series of products to show the functional properties of eggs.

Working in a group of three people, plan three different dishes that show coagulation, foam formation and denaturation of the protein in eggs (each person can make one of the dishes).

Write up your results as a group and then make a presentation to the rest of the class about which dish shows which property.

Proteins also have other characteristics.

When we are making bread, the dough has to be worked for a long time. We call this **kneading**. During the kneading process, the protein that is in the flour, called **gluten**, is developed and stretched. This is going to form the structure of the bread.

Figure 6.20 Products showing coagulation, foam formation and denaturation of egg protein

As the bread begins to rise, the gluten traps the carbon dioxide that is given off by the yeast, and the bread will rise to form the loaf, or bread rolls.

When the bread is cooked, the gluten sets, and forms the structure of the bread.

Figure 6.21 The gluten has stretched and then cooked to trap the air bubbles inside the bread

Practical activity

Work in a pair.
- One person makes a bread recipe using bread flour.
- The second person uses the same recipe but uses gluten-free flour.

Compare your results.

Explain why the gluten-free flour does not produce a well-risen loaf.

Denaturation can also happen when a protein has acid added. This can be in the form of vinegar or lemon juice, both of which are weak acids. This property is used in a **marinade**, where meat may be soaked in a flavoured liquid before cooking. The acid in the marinade begins to denature the protein, and makes the meat more tender before it is cooked. This is beneficial when barbecuing meat, as the method of cooking is quite fast, so does not allow the meat to tenderise much during cooking.

Figure 6.22 The acid in the marinade begins to denature the protein in the meat

Practical activity

Buy a piece of skirt beef or flank steak.
- Cut it in half.
- Soak half the steak in a marinade of your choice (overnight if you can).
- Pan-fry each piece and then cut each piece up.

Set up a blind tasting session.
- Ask the tasters to use sensory descriptors to describe the taste and texture of each piece.
- Get them to vote for the most tender and the most flavoursome.

KEY WORDS

Coagulation: where the protein sets as it cooks, forming a solid structure.

Foam formation or **aeration:** where whisking or beating an egg will result in the protein structure changing to trap small bubbles of air.

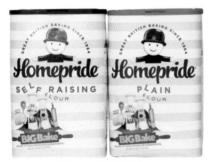

Figure 6.23 Different types of wheat flours available

Carbohydrates

In Chapter 2 we learnt that carbohydrates exist as starches and sugars.

Starch

One of the main starchy products we use for our recipes is flour.

Flour is usually made from wheat products (although it is now possible to buy gluten-free flour for coeliac sufferers).

Flour comes in several different types, and each one is used for a different function.

WJEC EDUQAS GCSE Food Preparation and Nutrition

Table 6.5 shows the different amounts of gluten found in wheat flour.

Type of wheat flour	Reason for use	Examples of recipes showing this function
Self-raising flour	This is a soft flour that contains baking powder as a raising agent. It contains less than 10% protein (gluten), so is not stretchy.	Cakes, biscuits and scones
Soft plain flour	This contains less than 10% gluten, so produces a soft dough that is not stretchy.	Shortcrust pastry Biscuits
Strong plain flour	This contains more than 10% of the protein gluten, so will produce a very stretchy dough when it is worked or kneaded, as the gluten will form stretchy strands to trap the carbon dioxide produced by the yeast.	Bread and bread products
Durum wheat flour	This is a special flour used to make pasta. It contains tough gluten, so will not stretch very well.	Fresh pasta

Table 6.5 Different types of wheat flour

Wheat flours also contain starch. The starch will not dissolve if it is mixed with cold water or another cold liquid such as milk.

When the mixed water and starch is heated, the starch granules start to absorb the water, or liquid, and they start to swell up. This causes the liquid to thicken. Some of the starch granules will then burst and release the starch, which then forms a gel. This process is called gelatinisation. (We discussed gelatinisation on page 105)

An example where this happens is when you are making a white sauce. If you look back at Figure 6.5, you can see a picture of a white sauce showing gelatinisation.

Gelatinisation also happens with oats, rice, pasta and potatoes.

The starch in flour products is also converted to sugar when heated. This is called **dextrinisation**.

This happens in dry heat, such as baking or grilling. You can see this when you toast a piece of bread, and the surface turns brown and crunchy, or when you cook a cake, and the surface goes golden brown.

Figure 6.24 Rice pudding showing gelatinisation

Figure 6.25 Bread showing dextrinisation as it turns into toast

Practical activity

Try to make a recipe using two different types of flour. For example, make some biscuits using soft plain flour and strong plain flour. Compare the results using a taste test. Get you testers to use sensory descriptors for their remarks.

Write up your results.

Practical activity

Make a rice pudding by using the boiling method on the hob.

The rice pudding will thicken as the rice grains absorb the milk and start to swell. This will show the effects of gelatinisation.

KEY WORDS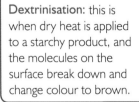

Dextrinisation: this is when dry heat is applied to a starchy product, and the molecules on the surface break down and change colour to brown.

Practical activity

Make the following quick biscuit recipe.

Biscuits

Ingredients

60 g caster sugar

120 g butter or margarine

180 g plain flour

Method

1 Cream the butter and sugar.

2 Add the flour and make into a dough.

3 Roll out the mixture and cut into biscuits.

4 Bake for ten minutes at 150°C in the oven. They will turn brown because of dextrinisation.

Figure 6.26 Different types of sugars available

Sugars

Sugar can be bought in various forms for different recipes.

● White sugar exists as granulated, caster, icing and preserving sugars.

● Brown sugars exist as demerara, soft brown and dark brown.

Function of sugar	Reason for use	Examples of recipes
To add sweetness and enhance the flavour of sweet foods such as fruits	Sugar is one of the five tastes that our taste buds recognise. Sweet foods are popular. Sugar is often added to ready-made meals to improve the flavour and encourage people to eat the product, enjoy it and therefore buy it again.	Cakes and biscuits Dessert recipes like fruit crumbles Sweet pastries
To add texture	Sugar is made of crystals, and depending on the size of the crystals, the texture of food will alter. Brown sugar can add a crunchy texture. Sugar also softens the gluten in flour when a recipe is baked. This gives a soft 'crumb' texture, like in a cake.	Cakes and biscuits Icing sugar gives a smooth coating for cakes Demerara sugar adds a crunchy topping for pies and pastries
To add colour	Figure 6.7 shows the different colours in caramelisation. The longer the sugars are cooked, and the higher the temperature, the darker the colour will be. This gives food a range of colours from light brown to black.	Cakes and biscuits Crème caramel Toffees and fudges
To trap air	If sugar and fat (butter or margarine) or sugar and egg are beaten or creamed together, the mixture will trap air. When the product is cooked it will help the product to rise.	All-in-one cake mixtures Creamed cakes e.g. Victoria sandwich

Table 6.6 The functions of sugar

Shortening

Shortening is the ability of fat to give a product a characteristically crumbly texture, such as in shortcrust pastry or biscuits. The fat coats the flour particles with a waterproof layer. This stops the gluten forming long strands and makes the final baked product have a soft melt-in-the-mouth texture (or mouthfeel) because the fat has shortened the gluten strands.

Aeration

The fat in a creamed recipe, such as a creamed cake, where the fat is creamed with caster sugar, will trap air. This will form a stable foam, which is then cooked. The trapped air will rise and be trapped by the gluten in the flour as the cake is cooked.

Plasticity

Different fats melt at different temperatures. This property is called **plasticity**. It means that each type of fat has a unique character. It is due to the mixture of combinations of chemicals in the fats, which are called triglycerides. Each one has its own melting point:

- Some products are created with lower melting points, so that they can spread straight from the fridge, such as 'Flora' and 'I Can't believe it's not Butter'.
- Other fats, such as butter, have a higher melting point, so will be solid when taken from the fridge, but will soften if left in a warm room.

Emulsification

Fats are **hydrophobic**. This means that they will not mix with water, but tend to form large globules when mixed with water, or a liquid. To allow them to mix with a liquid (**emulsification**), an **emulsifier** needs to be added. An example of this is when egg yolk is added to mayonnaise to allow the oil and vinegar to mix together.

Fats are solid at room temperatures. Examples are butter, margarine and lard.

Oils are liquid at room temperatures. Examples are vegetable oils and olive oil.

Fats and oils are very important in recipes. They add moisture, flavour and texture to foods.

Oils are used for cooking foods or as salad dressings or dips.

Figure 6.27 Butter, margarine and lard are all solid fats

Function of fats	Reason for use	Examples of recipes
To make cakes, biscuits and pastry Fats are sometimes called 'shortenings'	The fat creates a crumbly texture.	Shortcrust pastry Cakes Biscuits
To trap air	When beaten with sugar, fat will trap tiny bubbles of air. When the mixture is heated it will rise.	Creamed and all-in-one cake mixtures
To give flavour and moisture to recipes	The fat traps flavours. The fat also helps stop water evaporating from foods. This makes them last longer.	Fruit cakes Biscuits Some breads Fried foods
To make frostings and toppings for cakes and pastries	The fat is beaten with sugar to give a soft, flavourful topping.	Buttercream as a cake filling
Spreads on bread Putting on vegetables, such as potatoes, before serving	This adds flavour, and lubricates the food, making it easier to swallow.	Butter added to vegetables before serving
To add nutrients in the form of fat-soluble vitamins	Vitamin A and D are added to margarine by law.	

Table 6.7 Uses of fats in cooking

Function of oils	Reason for use	Examples of recipes
For roasting	This is when a food is baked in the oven, in oil. The temperature in the oven causes the water in the food to evaporate. This intensifies the flavour of the food. The oil adds moisture and helps to trap the flavour of the food.	Roast meat and poultry Roast vegetables such as potatoes, sweet potatoes, parsnips, butternut squash
For sautéing	This is when you cook a food in oil, over a low to medium heat. It brings out the flavour of a food, developing and releasing juices from the food.	Onions are slowly sautéed over a gentle heat to release sugars, which sweeten the onions, and also to caramelise the onions and turn them golden brown
For shallow-frying, stir-frying or deep-frying foods	The food is cooked and given a crispy coating and texture.	Shallow-frying: bacon, eggs, fish Stir-frying: vegetables, chicken, strip steak and seafood Deep-frying: battered fish, chips, doughnuts
To make salad dressings	This adds flavour and texture to foods.	Vinaigrette, mayonnaise
As a dip for breads	Adds flavour and moisture to a bread instead of using butter.	Oils can be flavoured with herbs and garlic

Table 6.8 Uses of oils in cooking

Stretch and challenge

1 Make a series of products using different forms of fats and oils. For example, select a cake recipe and make the same recipe using butter, margarine, lard and sunflower oil.
2 Compare the results.
3 Explain why the products have turned out differently. Use Tables 6.7 and 6.8, which explain the functions of fats and oils, to write up your results.

WJEC EDUQAS GCSE Food Preparation and Nutrition

Shortening: the ability of a fat to produce a characteristic crumbly texture to baked products.

Plasticity: the property of fat which allows different fats to melt at different temperatures.

Hydrophobic: a product, such as fat, that will not mix with water.

Emulsification: the property of fat that allows it to be mixed with a liquid after the addition of an emulsifier, such as egg yolk.

KEY POINTS: WORKING CHARACTERISTICS, FUNCTIONAL AND CHEMICAL PROPERTIES OF INGREDIENTS AND OILS

- Proteins exhibit coagulation, foam forming/aeration and denaturation as their properties.
- Weak acids, such as vinegar and lemon juice can also denature proteins.
- Gluten, the protein in flour, can stretch when worked, and form a structure which will trap air in a product such as bread.
- Carbohydrates are starches and sugars.
- Wheat flour contains gluten, which will form a structure when worked or kneaded.
- Starch will swell up when heated in a liquid, and will thicken the liquid.
- Starch molecules will break down and change colour when heated with dry heat.
- Sugars will sweeten and colour foods.
- Sugars will also change the texture of foods and trap air.
- Fats are either solid fats or oils.
- Fats are used to shorten gluten strands, make a softer food and to trap air.
- Fats are also used to give flavour, texture or add nutrients to a recipe.
- Oils are used for roasting, frying and sautéing foods.
- Oils are also used for salad dressing and dips.

Exam practice questions

1 Name **two** properties of eggs that show why they are used in cooking, and give examples of a recipe using each property. [4 marks]

2 Describe what happens to gluten during the kneading of bread dough. [3 marks]

3 Identify a recipe where it would be better to use a marinade before cooking the meat. [2 marks]

4 Suggest **two** recipes that use self-raising flour. [2 marks]

5 Give **two** functions of sugar in cooking. [2 marks]

6 Explain how caramelisation adds colour to the top of a cake. [3 marks]

7 Describe how fat helps create a crumbly texture to biscuits. Include reference to its shortening property. [4 marks]

8 What is meant by the term plasticity when referring to fats? Identify **two** different fats that show this property. [3 marks]

9 Explain how fat acts as an aerator in cake making. [3 marks]

10 Name **two** recipes that use butter or margarine, and explain the function of the fat in each recipe. [4 marks]

11 Explain how adding margarine to a recipe would increase the nutritional value of the final product. [2 marks]

12 Describe what happens when you sauté foods. [4 marks]

Figure 6.28 A cut apple showing oxidation due to enzymic browning

Chemical reactions in fruit and vegetables

When fruits and vegetables are cut or peeled, the surface of the fruit or vegetable is exposed to the air. The oxygen in the air reacts with the fruit or vegetable; this is called **oxidation**.

When the cells in the fruit or vegetable are cut, they release enzymes which react with oxygen, turning the fruit or vegetable brown. This is called **enzymic browning**.

To stop oxidation taking place, you can do one of the following:

● Squeeze lemon juice onto the cut surface of the fruit or vegetable.
● Cover the surface of the fruit or vegetable with cling film, to cut off the air to the cut surface.
● Cook or boil the fruit or vegetable straight after cutting.
● Blanch or steam the fruit or vegetable and freeze it.

KEY WORDS

Oxidation: the reaction of oxygen with the cut surface of a fruit or vegetable.

Enzymic browning: The release of enzymes from cut cells in fruit or vegetables, which react with oxygen and form brown discolourations on the product.

KEY POINTS: WORKING CHARACTERISTICS, FUNCTIONAL AND CHEMICAL PROPERTIES OF FRUITS AND VEGETABLES

● Cut surfaces of fruit and vegetables will discolour due to oxidation and enzymic browning.
● There are ways to prevent this occurring, including use of lemon juice, sealing in cling film, cooking the food or freezing the food after blanching.

Why some recipes do not succeed, and how to remedy situations

In this section, we will look at the ways that basic recipes may not turn out as expected and the reasons why these problems happen.

You can then use this information to correct the problem the next time you make the recipe.

Sometimes it is possible to remedy the situation, and sometimes it is not possible.

Sometimes you can change the product you are making into something else.

Cake making

Table 6.9 looks at different faults that can happen with a finished cake. It gives the reasons problems may occur and how you can avoid them.

Fault	Cause of fault	How to avoid or remedy the fault
The cake has sunk in the middle	1 The cake has not been cooked for long enough, and the middle was not set, so on removing it from the oven the middle sank. 2 Too much sugar was added to the recipe, softening the gluten too much so the cake collapsed. 3 Too much raising agent was added to the mixture so the gluten collapsed. 4 The oven door was opened during cooking, letting cold air in and causing the cake mixture to collapse before it was cooked.	1 Read the recipe to see how long the cake should be cooked for. Check it is ready by inserting a skewer in the centre of the cake – the skewer should come out clean. 2 Weigh the amount of sugar carefully. 3 Measure the amount of raising agent carefully. 4 Do not open the oven door until at least five minutes before the end of the cooking time specified in the recipe.
The cake has risen to a peak and the top has cracked	1 Too much mixture was put in the tin. 2 The oven temperature was too hot, and the top of the cake cooked before it had risen enough. 3 The cake was cooked too high up in the oven.	1 Only fill the tin three-quarters full to allow for rising. 2 Check the oven temperature matches what the recipe says. 3 Place the cake on the middle shelf of the oven.
The cake has a heavy texture	1 Too much liquid was used, and the cake mixture was too wet. 2 The oven temperature was too low. 3 Not enough raising agent was used. 4 During the mixing process the mixture curdled, and could not hold enough air.	1 Measure the liquid carefully. Only use what the recipe tells you. 2 Check the oven temperature matches what the recipe tells you. 3 Measure the amount of raising agent carefully. 4 If the mixture curdles when adding the egg, add a tablespoon of flour to the mixture and mix well.
The cake has an open and coarse texture	1 The flour was not mixed in properly. 2 Too much chemical raising agent was used.	1 Make sure all the flour is mixed in well. 2 Measure the amount of raising agent carefully.
The cake has risen unevenly	1 The mixture was not spread in the tin evenly. 2 The cake tin was put too close to the heat source in the oven, so rose more quickly on one side. 3 The oven shelf was not level.	1 Spread the mixture evenly over the whole tin. 2 Place the cake tin in the centre of the oven on the middle shelf. 3 Check the oven shelf is level before you put the cake tin into the oven.
The cake has a hard, sugary crust	1 A coarse sugar, such as granulated, was used, so did not dissolve properly. 2 Too much sugar was used. 3 The mixture was not creamed enough.	1 Use caster sugar for cake making. 2 Weigh the amount of sugar carefully and check it matches what the recipe requires. 3 Cream the butter or margarine and sugar until it is creamy and fluffy.
The cake has not risen	1 The mixture was not creamed enough, if the recipe was for a creamed mixture. 2 Insufficient raising agent, or the wrong type of flour was used. 3 If it was a whisked sponge, the eggs and sugar were not whisked until they were light and fluffy. 4 If it was a whisked sponge, the mixture was over-beaten when adding the flour, knocking all the air out.	1 Cream the butter or margarine and sugar until it is creamy and fluffy. 2 Measure out the raising agent carefully, and check the flour you are using is self-raising flour. 3 The eggs and sugar should be whisked until the mixture leaves a visible trail when the whisk is lifted. 4 Use a figure of eight movement when folding in the flour, so as to keep as much air in the mixture as possible.
The dried fruit or glacé cherries have sunk to the bottom of the cake	1 The cake mixture was too wet. 2 The fruit or cherries were washed and not dried properly before adding to the cake mixture.	1 Check the weight of flour you are adding is correct. Check you have used the correct-sized eggs. 2 Make sure you dry the fruit after washing, or coat it in the flour before you add it to the mixture.

Table 6.9 Common faults and causes of problems with cakes

Figure 6.29 A cake with a sunken middle

If the cake has already been cooked, it is not possible to correct any faults, but knowing why problems occurred can help you to avoid them next time you make the cake.

Practical activity

Work in a group of four people. Select a cake recipe. Each student makes the cake recipe, but makes the following changes:

- Person A follows the recipe exactly (this is the control).
- Person B uses plain flour instead of self-raising flour, and follows the recipe exactly.
- Person C uses granulated sugar instead of caster sugar, and follows the recipe exactly.
- Person D adds three extra tablespoons of sugar to the mixture during the creaming process, then follows the recipe exactly.

After cooking, look at Table 6.9, and see if the faults described there have occurred in your cakes.

Write up your results, describing what has happened.

Pastry making

Table 6.10 shows the common faults in pastry making, and the causes.

Fault	Cause of fault	How to avoid or remedy the fault
The pastry is sticky, soft and difficult to handle	1 Too much water was added. 2 A soft fat was used. 3 The mixture has been over-handled. 4 The pastry is too warm, so the fat has started to melt.	1 Measure the water carefully. 2 Use hard margarine or lard. 3 Handle the pastry as little as possible, and try to ensure your hands are not too hot. 4 If the pastry starts to get warm, wrap it in cling film and put it in the fridge for 20 minutes to relax.
The cooked pastry is hard and tough	1 Too much water was added. 2 The pastry was over-handled and the gluten has been over-developed. 3 Not enough fat was used. 4 The pastry was over-rolled.	1 Measure the water carefully. 2 Handle the pastry as little as possible, and try to ensure your hands are not too hot. 3 Weigh the amount of fat carefully. 4 Do not roll the pastry too much; try to get it the correct shape the first time. If it becomes warm, wrap it in cling film and place in the fridge for 20 minutes to relax.
The cooked pastry is dry and crumbly	1 Not enough liquid was used.	1 Measure the amount of liquid carefully.
The pastry shrinks when it is cooked	1 The pastry was stretched too much during rolling out. 2 The pastry was not allowed to relax before cooking.	1 Do not over-roll the pastry. 2 Put the rolled-out pastry in the fridge for 20 minutes to relax before cooking.
The pastry is oily and soft when cooked	1 The oven temperature was too low.	1 Always check that the oven temperature is correct, and the oven is allowed to heat up to the correct temperature before putting the pastry into the oven.

WJEC EDUQAS GCSE Food Preparation and Nutrition

The pastry is soft and crumbly	1 Not much water was used.	1 Measure out the amount of water carefully.
	2 Too much fat was used.	2 Weigh out the amount of fat carefully.
	3 Too much raising agent was used.	3 Measure out the amount of raising agent carefully.
The pastry blisters	1 The fat was not rubbed in correctly.	1 Make sure the mixture resembles breadcrumbs before the water is added.
	2 Too much water was used.	2 Measure the amount of water used carefully.
The pastry is pale	1 It was not baked for long enough.	1 Check the cooking time according to the recipe.
	2 The oven temperature was too low.	2 Always check that the oven temperature is correct, and the oven is allowed to heat up to the correct temperature before putting the pastry into the oven
The pastry is too dark	1 The oven temperature was too high.	1 Check the oven temperature is correct before putting the pastry in to cook.
	2 It was cooked for too long.	2 Always look at the dish five minutes before the end of the recommended cooking time, as some ovens cook faster than others.

Table 6.10 Common faults and causes of problems with pastry

Figure 6.30 A shrunken pastry case

Practical activity

1 Make some jam tarts.
2 Make a batch of shortcrust pastry. Divide it in two.
   - Roll one piece out and use a cutter to cut out and line some greased jam tart tins. Add the jam (this is your control).
   - Now take the other piece of pastry and roll it out several times before you cut out the pastry and make the jam tarts.
3 Cook both sets of jam tarts.
4 Compare the two results. Has the over-rolled pastry shrunk?
5 Compare the taste.

Bread making

Because bread has several stages during its making, there are various times when things can go wrong.

Table 6.11 lists some of the more common faults and causes.

Fault	Cause of fault	How to avoid or remedy the fault
Bread has not risen well and is coarse in texture	1 The yeast was killed before the bread was baked. 2 The yeast was out of date and did not activate. 3 The dough was left to prove for too long.	1 Only use warm water, if it is too hot it will kill the yeast. 2 If you are using dried yeast, check the use-by date. 3 The dough should be left until it doubles in size.
The bread has a dense texture	1 The wrong sort of flour was used; one with a high gluten content is needed, such as strong white, or bread flour. 2 Too much salt was added to the mixture, which will stop the yeast from working. 3 The yeast was killed before the bread was baked. 4 The dough was too dry, so could not expand properly. 5 The dough was left to prove for too long.	1 Check you are using strong plain, or bread flour. 2 Add the salt on the opposite side of the bowl to the yeast before you mix. 3 Only use warm water – if it is too hot it will kill the yeast. 4 Make sure enough water is added to give a soft, slightly sticky dough. Do not use too much flour on your board when kneading the dough. 5 The dough should be left until it doubles in size.
The dough collapses while baking	1 The dough was left to prove for too long.	1 The dough should be left until it doubles in size.
The bread is uneven in texture and has large holes	1 The bread was not kneaded enough after the first proving.	1 Make sure you knead the dough until it is elastic and springs back when you press it with your finger.

Table 6.11 Common faults and causes of problems with bread

Figure 6.31 Problems with bread

Sauces

There are three main methods to make a sauce:
- **Roux sauce**: the fat is melted in a pan, the flour is added and stirred and cooked for two minutes. The pan is removed from the heat and the milk is added little by little, stirring all the time. The pan is then returned to the heat, and stirred all the time until the sauce boils and thickens.
- **All-in-one method**: all the ingredients are put into the pan at the same time and the sauce is brought to the boil, stirring or whisking all the time until it thickens.
- **Blended method**: a little of the liquid is blended with cornflour or arrowroot. The rest of the liquid is heated and then added to the cornflour mix, stirring all the time. The whole mixture is then returned to the pan and brought to the boil, stirring, until it thickens.

They all have similar opportunities to go wrong.

Fault	Cause of fault	How to avoid or remedy the fault
The butter in the roux sauce burns before the flour is added	1 The pan was on too high a heat. 2 The butter was left too long before the flour was added.	1 Melt the butter on a gentle heat. 2 As soon as the butter melts, add the flour and stir well.
The roux sauce tastes floury	1 The flour was not cooked long enough with the butter before the milk was added. The sauce was not cooked for a couple of minutes once boiled.	1 Cook the butter and flour mixture (the roux), for three to four minutes, stirring all the time, to cook the flour.
The sauce mixture is lumpy before being brought to the boil	1 The milk was added too quickly, and the mixture was not whisked or stirred enough during the addition of the milk.	1 Add the milk gradually, stirring well between each addition.
The sauce is catching and burning at the bottom	1 The heat is too high. 2 You are not getting into the corners of the pan, and the flour is catching underneath. 3 You are not whisking or stirring fast enough.	1 Have the heat on a medium flame for gas cookers, or a middle number for electric cookers. 2 Make sure you are reaching into all the edges of the pan while you are stirring the sauce. 3 The sauce should be stirred vigorously, particularly when it is beginning to thicken.
The sauce is lumpy when it has thickened	1 You did not whisk or stir quickly enough when the mixture was coming to the boil and thickening.	1 The sauce should be stirred vigorously, particularly when it is beginning to thicken.

Table 6.12 Faults and causes of problems with sauces

Emulsion sauces

These sauces use a mixture of oil and vinegar, with the addition of an emulsifier, which prevent the oil and vinegar separating.

Sauce	Fault	Cause of fault	Remedy
Mayonnaise	1 The mayonnaise fails to thicken. 2 The mayonnaise becomes oily on the surface.	1 The oil has been added too quickly, so it never gets dispersed. 2 Water has evaporated from the mixture, giving the oil droplets a chance to coalesce.	1 Beat a fresh egg yolk with a tablespoon of water and/or lemon juice in a clean bowl, and slowly whisk into the broken sauce. 2 Whisk in a spoonful of water.
Hollandaise	1 The sauce is lumpy and thin. 2 The sauce has separated while being kept warm.	1 The egg yolk has overcooked. 2 The sauce has become too hot, causing the butterfat to leak.	1 Strain out the lumps and whisk the hot broken sauce into another gently heated egg yolk in clean bowl. 2 Take it off the heat and whisk it vigorously, or briefly re-emulsify it in a blender.
Vinaigrette	1 The oil and vinegar have separated.	1 The simplest vinaigrettes do not contain emulsifiers like mustard, so the oil and vinegar separate unless they are being actively mixed.	1 Whisk the broken vinaigrette in a bowl or shake it vigorously in a closed jar and pour it over the food immediately, while it's still in motion. A separated mustard vinaigrette is fixed in the same way.

Table 6.13 Faults, causes and remedies for emulsion sauces

Practical activity

Work with another student.

Make a roux sauce.
- One student follows the recipe exactly.
- The other student follows the recipe, but only stirs the sauce occasionally while the sauce is coming to the boil.

Compare your results. Write up what happened.

KEY POINTS: WHY PARTICULAR RESULTS MAY NOT BE ACHIEVED AND HOW TO REMEDY SITUATIONS

- It is very important to follow the steps in recipes, from the weighing of ingredients to the making and cooking of the dish.
- If a recipe goes wrong, it is sometimes possible to remedy it, but it is much better to avoid the problem in the first place.
- If you make a mistake, it is often easier to start from the beginning.

Exam practice questions

1   List **three** possible causes of a cake sinking in the middle during baking.   [3 marks]

2   If your pastry becomes too warm, and the fat is beginning to melt, what should you do to allow it to become easier to handle?   [2 marks]

3   You have used a packet of dried yeast for your bread. The bread has not risen properly. Name **two** possible causes for this.   [2 marks]

CHAPTER 7
Food spoilage

Learning objectives

In this chapter you will learn:

- how to store foods correctly and the different methods used for food storage
- how to prevent and control methods for enzyme action, mould growth and yeast production
- the different signs of food spoilage
- the role of temperature, pH, moisture and time in the control of bacteria
- the different types of bacterial cross-contamination and how to prevent it
- the different methods of food preservation
- the signs and symptoms of salmonella, campylobacter, e-coli and staphylococcus food poisoning
- the impact of food wastage, including the effect on the environment and the financial implications of waste.

Storing foods correctly

Food must be stored correctly so that it is safe to consume. Storing food correctly can help prevent **food spoilage**. There are different ways of storing food; the method chosen will depend on the type of food that is being stored.

Refrigeration of food/cold storage

When foods are refrigerated they provide less risk of **food poisoning**. The cold temperature slows down any bacterial growth and helps to prevent bacteria activity.

There are certain foods that need to be kept in the refrigerator, for example fresh foods such as milk, meat, dairy produce, ready meals and some desserts. Some foods state 'keep refrigerated' on the label, for example, salads, some fruits and vegetables, and ready-prepared items like coleslaw and dips. Other foods state 'refrigerate after opening', for example, UHT milk, jams, chutneys and sauces.

When storing food in the fridge, it is important that you store it in the correct place to prevent bacteria from raw foods contaminating cooked foods and ready-to-eat items.

Rules for storing food in the refrigerator:

- Keep your fridge temperature at 5°C or below.
- Do not overload the fridge; this can prevent circulation of cool air which could result in food not being kept cold enough.
- Avoid putting hot food in the fridge – allow it to cool first.
- Remove any food that has gone past its use-by date.
- Only store food for its recommended time.

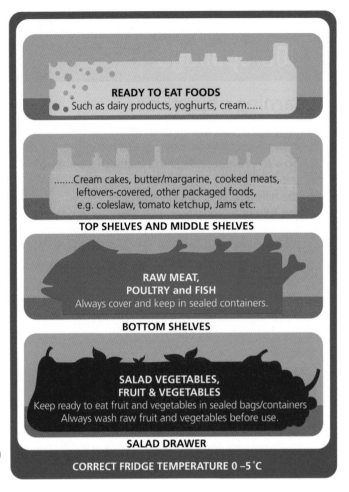

READY TO EAT FOODS
Such as dairy products, yoghurts, cream.....

.......Cream cakes, butter/margarine, cooked meats,
leftovers-covered, other packaged foods,
e.g. coleslaw, tomato ketchup, Jams etc.

TOP SHELVES AND MIDDLE SHELVES

**RAW MEAT,
POULTRY and FISH**
Always cover and keep in sealed containers.

BOTTOM SHELVES

**SALAD VEGETABLES,
FRUIT & VEGETABLES**
Keep ready to eat fruit and vegetables in sealed bags/containers
Always wash raw fruit and vegetables before use.

SALAD DRAWER

CORRECT FRIDGE TEMPERATURE 0 –5°C

Figure 7.1 How to store food in the refrigerator

- Wrap food or put in covered containers before storing.
- Clean the fridge regularly.
- Avoid opening the door unnecessarily to prevent warm air entering.

Freezing of food

When food is frozen, the **shelf life** of the food is extended. The very cold temperature stops the growth of micro-organisms and **enzymes** that cause food spoilage. The length of time the food can be stored for will depend upon the star rating of the freezer.

- A one-star freezer (*) is normally an ice shelf or freezer box in a fridge. The temperature is –6°C and it can store pre-frozen food for up to a week.
- A two-star freezer (**) temperature is –12°C and it can store pre-frozen food for up to 1 month.
- A three-star freezer (***) temperature is –18°C and can store pre-frozen food for 3 to 12 months. Most domestic freezers are this temperature.

When food is frozen, the water content in the food becomes solid; this prevents any multiplying of bacteria. When food is thawed, its structure can be damaged and break down; the colour, texture, flavour and nutritional value of the food can also be affected. Bacteria activity will start when the food has reached a suitable temperature; this is why it is important that thawed foods are not refrozen.

Rules for storing food in the freezer:

- Follow recipes which give clear instructions on how to prepare foods like fruits and vegetables from frozen.
- Wrap food well so that it is completely covered and free from air. Freeze your frozen food in appropriate containers, e.g. freezer bags and airtight containers. Badly packaged food can lead to freezer burn which occurs when air meets frozen food.
- Label packaged food with the contents and date.
- Do not overload the freezer as this can affect the temperature.
- Keep a freezer thermometer in the freezer and check the temperature regularly.
- Store new frozen food at the back or near the bottom and move older ones near the top and front so that they are used first.
- Check the best-before date to avoid food going past its quality peak and to ensure it is safe to use.

Dry storage of food

Dry storage is used for foods that do not need to be stored in a refrigerator (chilled) or freezer (frozen). Foods generally stored in a dry storage area include canned goods, flour, sugar, rice and cereal. Some vegetables such as potatoes and onions can also be stored this way.

It is important that the food is stored so that it is safe to eat and that the food maintains its quality.

Foods stored using this method:

- can be kept at temperatures between 10°C and 21°C, however the cooler the better
- should be kept in a dry, well-ventilated area
- should be kept out of direct sunlight
- should be kept in the original packaging, or stored in airtight containers that are clearly labelled
- should be rotated so that the oldest food is used first
- should be stored off the floor to avoid **contamination** from insects and vermin
- should be dated to ensure they are used within the recommended time.

Exam practice questions

1 State the correct temperature for a refrigerator. [1 mark]
2 List **three** rules to follow when storing foods in a refrigerator. [3 marks]
3 Describe the correct way to store raw meat and poultry in the refrigerator. [2 marks]
4 Explain the differences between a one star (*), two star (**) and three star (***) freezer. [6 marks]
5 Describe the effect freezing can have on some foods. [3 marks]

Date marks and labelling

When food products are bought they will come with a **date mark**. This could be a 'use-by' date or a 'best-before' date.

- **Use-by date:** this means the product is safe to consume up to and including this date.
- **Best-before date:** this is a recommendation that the product is used before this date, when it is of its best quality. After this date the product will still be safe to use, but may have deteriorated in quality.

Date marks make sure the buyer is aware of the shelf life of the product, so they know how long they can keep the food for and for how long the food will be safe to eat.

When storing foods in the refrigerator or freezer at home, it is important that it is clearly dated. This will ensure that the food is used within its appropriate recommended storage time and that the food will be safe to consume.

When storing foods in the freezer they should be clearly labelled with the contents so that you are aware of the contents of different containers of food and do not get them mixed up when removing them. Important information such as how they have been prepared, for example if the vegetables have been blanched, should also be included.

When using food from the refrigerator or freezer the **First in First Out (FIFO)** rule should apply, so that older items are used first. This can also prevent food wastage.

> **KEY WORDS**
>
> **Date mark:** date labels added to products. Use-by date and best-before dates are both date marks.
>
> **Use-by date:** the product is safe to consume up to and including this date.
>
> **Best-before date:** the date by which it is recommended that the product is used, when it is of its best quality.
>
> **First in First Out:** a rule used when storing food, so that older items are used first.

Exam practice questions

1 Explain the differences between a 'use-by' date and a 'best-before' date. [2 marks]

2 Explain why it is important that food is clearly dated when storing in a freezer. [2 marks]

Growth conditions, prevention and control methods for enzyme action, mould growth, yeast production and bacteria

Growth conditions

The conditions required to enable micro-organisms to grow and spoil food can vary depending on the type of micro-organism. Micro-organisms can make food unsafe to eat:

- by simply being in the food
- by producing waste products in the food
- by producing toxins or poisons in the food.

The main micro-organisms are:

- **Yeast** – yeasts are found in soil, in the air and on the skin of some fruits such as plums. They are one-celled organisms which come from the group of organisms called fungi. Yeasts reproduce by budding; this means they multiply and the one cell divides into two.
- **Moulds** – moulds are a type of fungi; they are from the same family as mushrooms. They can be blue/green, white or black in colour. Moulds reproduce by producing spores, which will travel in the air and land on food. The spores will grow if the conditions are correct.
- **Bacteria** – bacteria are extremely small single-celled organisms. They can only be seen under a microscope. Bacteria are found everywhere including in air, food, water and on animals and humans. Bacteria are able to reproduce very quickly by dividing into two about every 20 minutes if the conditions are correct; this means millions can be produced in 24 hours.

Figure 7.2 Yeast budding

The conditions required to enable growth of the micro-organisms are shown in the table below.

Micro-organism	Growth conditions
Yeast	Warmth
	Water/moist conditions
	A food supply (e.g. sugar)
	Time to be able to reproduce
Moulds	Warmth
	Moist conditions – room temperature is ideal
	A food supply
Bacteria	A food supply
	Moisture
	A suitable temperature; 37°C is the best; however bacteria will grow within the temperature range of 5°C to 63°C
	Oxygen can also be a condition of growth

Table 7.1 Growth conditions for micro-organisms

Figure 7.3 Mould on a loaf of bread

Figure 7.4 Enzymic browning
– an apple turning brown when
cut and exposed to oxygen

Enzyme action

Enzymes are protein molecules that control chemical reactions in food. They are found in plants and animals and are inactive until food is either harvested or slaughtered.

Once enzymes are activated, they can cause unwanted or undesirable changes in foods such as a colour change. An example of this would be the browning of the surface of a cut apple when it is exposed to oxygen in the air. This is due to the tissues of the fruit being broken down. This is often referred to as enzymic browning.

Enzymes also turn starch into sugars in fruits, which can cause them to ripen and over ripen.

How to prevent and control enzyme action

Enzyme action can be prevented and controlled by:
- a heat treatment such as **blanching** – when preparing vegetables such as potatoes that have been peeled and sliced, if you plunge the potato slices into boiling water the intense temperature will stop the potatoes being affected by the enzymes; it inactivates them
- **acidic conditions** – when preparing apples, if you coat or dip cut apple pieces in lemon juice, which contains citric acid, this can prevent the apples from browning.

Mould growth

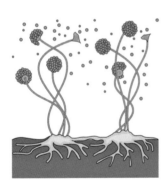

Figure 7.5 Mould growing

Moulds are very small plants that can grow on many types of foods. They have threads that grow up and roots that grow down into the food and take away all of the nutrients in the food.

Mould growth could be described as being fuzzy in appearance; it is of different shapes. It can be seen on the surfaces of foods by the eye. Foods are not safe to eat if they have mould on the outside but they look normal on the inside. Moulds can produce harmful substances that can spread into the food.

Some moulds that are considered to be harmless are used in the production of blue-veined cheeses, for example Danish Blue. They are used to add flavour to the foods.

How to prevent and control mould growth

Moulds can grow on foods that can be slightly acid, alkaline, sweet, salty, moist or dry. They are very productive in warm, moist conditions. To prevent mould growth:
- store suitable foods in a chilled place such as a refrigerator
- foods that do not need refrigeration should be stored in a cool, dry place
- heat foods to a very high temperature (e.g. above 100°C) to prevent the production of heat-resistant spores
- store foods, if suitable, in very acidic conditions.

Figure 7.6 Danish Blue cheese

Yeast production

Yeast is widely used in the production of bread and some alcoholic drinks such as beer. Yeast cells do not need oxygen to reproduce; this means they are **anaerobic**. Yeast cells will reproduce provided they have the following conditions:

- water or liquid
- a warm environment
- a food supply
- a period of time.

How to prevent and control yeast production

- Keep yeasts cold as they are inactive or **dormant** in cold temperatures.
- Yeast causes high-sugar foods such as jam and fruits to spoil; this can be prevented by storing in a cold environment.
- Keep dried or fresh yeast away from moisture or liquid as this can start activation.
- Yeasts are killed at high temperatures (e.g. 100°C).

Bacteria

There are many different forms of bacteria, all of which have different shapes. If bacteria have the correct conditions they will rapidly reproduce. Bacteria produce **toxins** which can be very harmful to humans, causing food poisoning.

How to prevent and control bacteria production

- Store food in refrigerators that are kept at temperatures between 0°C and 5°C; this will be too cold for the bacteria and should prevent bacteria production.
- Food that has been cooked to eat later, or is left over after a meal should be cooled rapidly to 0°C to 5°C to prevent harmful bacteria from being active as they love warm, moist conditions. Cooling of these foods should normally be within 90 minutes.
- Leftover foods, for example a meat-based bolognaise sauce, should be eaten within 24 hours and must only be reheated once. Reheating will activate the bacteria and cause multiplying.
- Bacteria can be destroyed or prevented from multiplying by the use of a strong acid for example vinegar, or a high amount of sugar and salt.

KEY WORDS

Yeasts: one-celled organisms which come from the group of organisms called fungi. Yeasts reproduce by budding; this means they multiply and the one cell divides into two.

Moulds: a type of fungi. Moulds reproduce by producing spores, which will travel in the air and land on food.

Bacteria: extremely small single-celled organisms that can only be seen under a microscope. Bacteria are found everywhere including in air, food, water and on animals and humans. Bacteria are able to reproduce very quickly by dividing into two about every 20 minutes if the conditions are correct.

Anaerobic: do not need oxygen.

Dormant: not active, asleep but not killed.

Toxins: poisons that can cause illness.

KEY POINTS: GROWTH CONDITIONS, PREVENTION AND CONTROL METHODS FOR ENZYME ACTION, MOULD GROWTH, YEAST PRODUCTION AND BACTERIA

- Yeasts, moulds and bacteria all need certain conditions to enable them to grow/reproduce (e.g. food, warmth and moisture).
- Enzymes can cause undesirable changes in foods such as causing fruits to ripen quickly. Enzymes can be controlled by heat and the use of an acid.
- Mould growth can be seen on the surface of foods. Foods that have moulds growing on them should not be eaten. Some moulds are used in the production of cheeses and are considered harmless.
- Yeasts require liquid, warmth, food and time to reproduce. They are dormant in cold temperatures and killed at high temperatures.
- Bacteria produce toxins which can cause food poisoning. Foods must be stored correctly to prevent bacteria reproducing in them; temperature plays a vital role. Acids or high concentrations of sugar or salt can prevent reproduction of bacteria.

Exam practice questions

1 Name the **three** different types of micro-organisms.
[3 marks]

2 Which micro-organism can only be seen under a microscope?
[1 mark]

3 Name the micro-organism which is found on the skin of some fruits.
[1 mark]

4 Identify the **three** conditions moulds need to reproduce.
[3 marks]

5 Explain what is meant by an enzyme. [2 marks]

6 Describe how enzymes can change the appearance of a food.
[2 marks]

7 Explain what enzymic browning is and give an example of a food affected by it.
[3 marks]

8 Describe **one** method used to prevent enzyme action.
[3 marks]

9 Explain the differences between mould and yeast contamination.
[2 marks]

10 List the **four** conditions needed by yeasts to become active.
[4 marks]

11 Why should leftover foods only be reheated once?
[1 mark]

12 Explain why toxins can be harmful to humans. [2 marks]

Signs of food spoilage

Fresh food can spoil quite quickly during storage. During this time changes could be noticed in the texture, flavour or colour of food. Changes in food are generally caused by micro-organisms such as bacteria, moulds and yeasts. Enzymes will also cause food to spoil. Food can also spoil during storage due to the natural decaying of the food itself. Food that has spoiled is often referred to as food that has 'gone off'.

There are many signs that can help determine if food has spoiled. These include:

- a sour smell or taste
- areas of mould on the food
- a slimy feel to the surface of the food
- a loss of moisture leading to wrinkled foods that can be discoloured
- foods looking and feeling over-dry or over-wet.

Food spoilage could also be caused by:

- contamination of the food
- poor handling of the food
- incorrect storage or poor storage of the food.

KEY POINTS: SIGNS OF FOOD SPOILAGE

- Food spoilage can be caused by micro-organisms, enzymes and the natural decaying of the food.
- The smell, texture and appearance of foods can help determine if foods have spoiled.

Exam practice questions

1 State **three** causes of food spoilage. [3 marks]

2 Identify three signs that could indicate if food has spoiled. [3 marks]

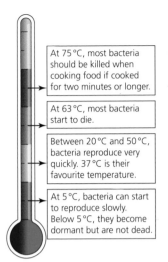

At 75°C, most bacteria should be killed when cooking food if cooked for two minutes or longer.

At 63°C, most bacteria start to die.

Between 20°C and 50°C, bacteria reproduce very quickly. 37°C is their favourite temperature.

At 5°C, bacteria can start to reproduce slowly. Below 5°C, they become dormant but are not dead.

Figure 7.7 Different temperatures and bacteria growth

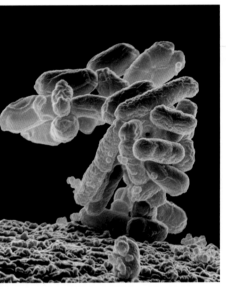

Figure 7.8 Bacterial cells multiplying

The role of temperature, time, pH and moisture in the control of bacteria

Temperature and time are two crucial conditions that must be controlled to prevent the growth or reproduction of bacteria.

Temperature

The temperature bacteria like best is 37°C; this is body temperature. They are most active in the temperature range of 5°C – 63°C. This temperature range is referred to as the **danger zone.**

In different temperatures bacteria will reproduce at different speeds, from rapidly to very slowly.

Time

Bacteria have the ability to reproduce very quickly; they do this by continually dividing in two. This can be as often as every 10–20 minutes. One bacteria cell can develop into a whole colony of bacteria within 12 hours. By 24 hours, millions of bacteria will have been produced. Once bacteria have other required conditions such as food, warmth and moisture, giving them time is lethal.

It is therefore very important that food storage is at the correct temperature to prevent bacteria from reproducing and that foods that are of a particular risk – for example, raw meats and fish, cooked meats and fish, and dairy products – are stored for the correct length of time in the correct conditions.

pH

A pH scale is used to measure how acid or alkaline a substance is. Acidic foods have a pH below 7 and alkaline foods have a pH above 7. Bacteria prefer neutral conditions, such as pH7, and they are able to grow rapidly in these neutral conditions. Many pathogenic bacteria are not able to grow in acidic or alkaline environments.

Moisture

Soups, sauces and gravies are also considered to be high-risk foods due to their moisture content and high protein content (this is dependent on the ingredients used to make them). Bacteria thrive in these types of environments.

Exam practice questions

1 Describe how bacteria multiply. [2 marks]
2 Name **three** foods that bacteria like to multiply in. [3 marks]
3 Explain why gravy is considered a high-risk food. [2 marks]
4 Explain what is meant by the 'danger zone'. [2 marks]

Types of bacterial cross-contamination and their prevention

When food is being stored, prepared and cooked there are many opportunities for **cross-contamination** to take place. Cross-contamination is when micro-organisms such as bacteria are transferred from a source such as raw food to another source such as cooked or ready-to-eat food. This can then lead to food poisoning.

The most common causes of bacteria cross-contamination include:
- raw food coming into contact with cooked foods – an example of this could be having raw meat on the same chopping board as cooked chicken and they touch
- raw meat juices dripping onto cooked meat or another high-risk food – an example of this could be when storing steaks in the refrigerator and they are stored incorrectly at the top of the refrigerator allowing blood to drip down
- hands transferring bacteria from one food to another – an example of this could be a food handler touching raw food and then not washing hands thoroughly before moving on to the next task
- equipment transferring bacteria from one food to another – an example of this could be using a knife to cut raw chicken into pieces and then using the same knife to slice some cooked ham
- dirt from vegetables coming into contact with high-risk foods – an example of this could be using the same chopping board to slice some unwashed leeks and then using the same board and knife without washing them to slice up some cooked roast beef.

How to prevent bacterial cross-contamination

Bacterial cross-contamination can be controlled and prevented by ensuring good practices and following basic rules such as:
- washing hands before cooking as well as after touching raw meat, fish and dirty vegetables; hands should also be washed after using the toilet and after blowing your nose
- covering and storing raw meats on the bottom shelf of the refrigerator on a suitable plate or tray and away from cooked meats to avoid any juices dripping onto other foods
- using clean equipment for each task, e.g. when preparing raw foods, cooked foods, or when preparing foods that do not require cooking
- when cooking raw meat, for example frying steaks, putting all of the raw meat in to cook at the same time, not adding raw meat to the pan when the other pieces of meat are cooked
- using clean dish cloths and tea towels when preparing foods to avoid bacteria being transferred from dirty cloths to equipment pieces and then onto foods to be eaten.

KEY WORDS 🔑

Cross-contamination: the transferring of bacteria from one source to another.

KEY POINTS: CROSS-CONTAMINATION 🎯

- Bacterial cross-contamination can lead to food poisoning.
- Bacterial cross-contamination can take place during food storage, preparation, cooking and food handling.
- There are basic rules that can help prevent bacterial cross-contamination. These include good personal hygiene, correct storage of foods, the use of clean equipment and the correct handling and cooking of food.

Exam practice questions

1 Give an example of when cross-contamination can take place during the preparation of food. **[3 marks]**

2 Explain how raw meat should be stored in a fridge and why the positioning of the meat is so important. **[3 marks]**

3 Explain why it is important to put all raw pieces of meat into a frying pan at the same time. **[2 marks]**

Preservation

Food is preserved to prevent it from spoiling or 'going off'. When you preserve food you are preventing the growth of micro-organisms such as bacteria and moulds and trying to delay the decaying of food. Preservation aims to extend the shelf life of a food product.

Freezing

When foods are frozen the water content in them will become solid as they freeze. The cold temperature of the food means that the growth rate of micro-organisms is reduced, as they are inactive in these conditions. With the water being frozen there is no liquid available, which is also a requirement for growth.

Upon defrosting or thawing, micro-organisms will slowly become active and start to multiply again. Once food has defrosted or thawed it should not be refrozen.

When freezing foods it is better to freeze them at the peak of their quality as they will retain their flavour compared to foods that are frozen when they are not so fresh.

Some foods such as vegetables can be blanched before freezing; this will inactivate enzymes found in the foods, and the hot temperature used when blanching the vegetables will reduce the number of bacteria present.

The temperature a domestic freezer should be kept at is $-18°C$.

Freezer burn

Foods for freezing must be packaged correctly to avoid freezer burn. Freezer burn will dehydrate the food, resulting in a colour change and have an impact on the flavour and texture.

Some foods do not freeze well; this can be due to the large proportion of water that they contain. Examples of foods that do not freeze well include bananas (these will turn black) and single cream (this will separate).

Figure 7.9 Freezer burn on a hamburger

Pickling

Foods that are pickled are completely covered in vinegar and are generally stored in glass jars.

Examples of pickled foods include:
- vegetables – pickled onions, gherkins
- fish – pickled herrings
- protein-based foods – eggs.

The pickling vinegar must contain acetic acid to prevent the growth of micro-organisms. The vinegar will act as a preserving agent.

When preparing foods such as vegetables for pickling, they are normally treated with salt first. This will help extract some of the water from the vegetable and prevent the growth of some bacteria.

Figure 7.10 Freezer burn showing colour change on a piece of meat

Spices are also used when pickling different foods such as fruits to provide flavour. Some fruits will be cooked gently in spiced vinegar, and sugar will be added before placing the fruits in the jars. The heat can help kill any micro-organisms and the sugar will give a sweeter flavour as well as help to prevent the growth of micro-organisms. Enzymic action can also be reduced. Once the jars are sealed they are often left to mature, which can take several weeks.

Figure 7.11 A jar of pickled onions

Jam making

Making jam is a good way to preserve fruit when it is in season and at its cheapest. Jams are made from fruit, sugar and water.

During the making process the ingredients are heated to a very high temperature; this destroys any micro-organisms present – the main one is normally yeast in the fruit. The level of sugar used in jam making is high; this acts as a preserving agent and helps to prevent the growth of micro-organisms.

The jars used to store the jam in must be sterilised before use and the jam has to be placed in the sterilised jars whilst it is still very hot. The jars then must be sealed immediately to prevent contamination from micro-organisms.

Bottling

To bottle foods, you cook foods such as fruits or vegetables in sealed sterile jars. The cooking process involves heating the jars and boiling them for a specific time to kill any harmful micro-organisms that may be in the food inside the jars.

The foods are either placed in a salt solution which is called brine or a sugar solution referred to as syrup. Both of these solutions can help prevent the growth of micro-organisms. The exclusion of air when the jars are sealed can also help preserve the foods.

When the jars are cooled the food inside will be preserved for several months provided the jars are not opened.

Vacuum packing

Vacuum packing is a method of increasing the shelf life of a food product by removing the oxygen. The method involves the food product being placed in an airtight pack, the air is then sucked out and the package sealed. This means the food is kept in anaerobic conditions.

This process has to be carefully controlled by chilling, as the lack of oxygen can mean that other pathogenic micro-organisms will grow. By removing air from around the food product, the levels of oxygen in the packaging are reduced, which will restrict the ability of oxygen-breathing micro-organisms to grow and spoil the product.

Examples of vacuum packaged foods:
- meat joints and meat products
- fish and fish products
- fresh pasta
- vegetables.

Figure 7.12 Vacuum packaged meat

> ### KEY POINTS: PRESERVATION
> - Preservation aims to prevent the growth of micro-organisms and to extend the shelf life of foods.
> - The main methods used to preserve foods are freezing, pickling, jam making, bottling and vacuum packing.
> - Freezing reduces the temperature to prevent micro-organism growth.
> - Pickling can include applying heat and putting foods in acid conditions to prevent micro-organism growth.
> - Jam making and bottling increases the temperature to prevent the growth of micro-organisms. High sugar and salt solutions are also used.
> - Vacuum packing removes the oxygen to extend the shelf life of foods.

Exam practice questions

1 Explain the process of blanching.　　[3 marks]　　2 Explain what the term 'freezer burn' means.　　[2 marks]

Signs and symptoms of food poisoning

Food poisoning is an illness that is caused by the consuming of food or water that has been contaminated by specific pathogenic bacteria or their toxins, or by other harmful micro-organisms.

There are many reasons why food poisoning can occur. Some of these include:
- poor hygiene practices including handling of food and cleaning of food areas
- incorrect temperature control when reheating foods and keeping foods hot
- incorrect storage of foods
- incorrect preparation and cooking of foods
- reheating of ready meals in the home at incorrect temperatures and for insufficient times.

Bacterial contamination

The most common cause of food poisoning is bacterial contamination. Bacteria need four requirements to be able to reproduce:
- **Food** – bacteria need nutrients to be able to grow, these are found in foods particularly high-risk foods which bacteria like to grow on.
- **Warmth** – bacteria like warm temperatures, particularly 37°C. Bacteria will mostly reproduce between the temperatures of 5°C and 63°C.
- **Moisture** – bacteria need moisture to reproduce. Foods such as gravy are ideal. Without the moisture they cannot reproduce, so gravy powder would be safe.
- **Time** – bacteria need time to reproduce. As they reproduce, they divide and then multiply so one bacterium becomes two and two become four. This method is called binary fission.

With all the requirements above, bacteria will reproduce every 10 to 20 minutes.

If you take away one of more of these requirements, then the bacteria could be slowed down or prevented from growing. This could help reduce the risk of food poisoning.

There are different types of pathogenic bacteria that will cause different food poisoning symptoms. This type of bacteria rarely change the smell, appearance or taste of the food. Some examples of food poisoning are shown in the table below.

Type of bacteria	Food poisoning symptoms	Food sources	Where the bacteria is found
Salmonella	Fever, headache, abdominal pain, diarrhoea, nausea, vomiting	Poultry, chicken and other types of meat, raw eggs or undercooked eggs, egg products and some dairy foods	Raw foods, dirty water, pests and transferred from people
Campylobacter	Diarrhoea (this may be bloody), stomach cramps, fever, and vomiting	Raw or undercooked meat, particularly poultry, unpasteurised milk and untreated water	Animals, untreated water, sewage, pests
E-coli	Nausea or vomiting Severe abdominal cramps Watery or very bloody diarrhoea Fatigue Fever	Raw meat, meat dishes and products, e.g. cooked meat and gravy, unpasteurised juices, cheeses and milk, raw fish and oysters	Dirty water, raw meat that has come in contact with bacteria, and waste from animal intestines during processing, human waste
Staphylococcus	Abdominal pains Vomiting, diarrhoea, can cause people to collapse	Milk, cream, cooked meats and meat products, egg products, salads such as egg, tuna, chicken, some dairy products	On people (skin, nose, mouth, hair) from cuts, burns, scratches and skin infections, raw, untreated milk from cows and goats

Table 7.2 Causes and symptoms of food poisoning

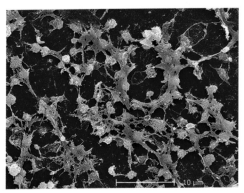

Figure 7.13 Staphylococcus bacteria

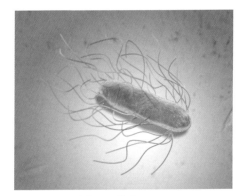

Figure 7.14 Salmonella bacteria

KEY POINTS: SIGNS AND SYMPTOMS OF FOOD POISONING

- Food poisoning bacteria rarely change the smell, appearance or taste of food.
- Salmonella, campylobacter, e-coli and staphylococcus are all types of pathogenic bacteria that cause food poisoning.
- Staphylococcus is the bacteria that is found on human skin, in the nose and mouth, as well from cuts and skin infections.
- Two common symptoms of food poisoning are diarrhoea and vomiting.

Exam practice questions

1 Give **two** reasons why food poisoning occurs. [2 marks]
2 Name **two** symptoms of salmonella food poisoning. [2 marks]

3 Describe in detail **two** of the requirements needed by bacteria to be able reproduce. [4 marks]

Food wastage

As a country we generate a large amount of food wastage. According to the 'Love Food Hate Waste' website we throw away 7 million tonnes of food and drink from our homes every year in the UK, and more than half of this is food and drink we could have eaten.

The main types of foods that are thrown away as waste include fresh vegetables and salad, fresh fruit, bread, cakes, prepared foods such as pasta and rice, as well as meat-based meals and takeaways. On many occasions whole, unopened packs of these items are disposed of.

There are many reasons why we waste so much food. These include:

- The food has gone past its date for using.
- The food has an unpleasant smell, look or taste.
- The food has gone mouldy.
- Too much food was cooked leaving waste.
- Food has been left on the plate – this could be due to large portion size, dislike of food, or not feeling hungry.

Effect on the environment

To try to reduce the impact on the environment that food wastage can have, many areas of the UK have, as part of their recycling schemes, introduced food waste bins. Households are encouraged to place waste or leftover foods into the bins; the waste is then turned into compost or used to help generate energy.

When food is thrown away through the normal rubbish disposal system, it is taken to a landfill site where it will rot and as a result produces methane, a greenhouse gas which is much more harmful than carbon dioxide.

Figure 7.15 A landfill overflowing with edible food waste

Financial implications of waste

When we waste food, not only are we throwing away food that could have been eaten, we are also throwing away money. 'Love Food Hate Waste' states that wasting food costs the average household £470 a year, rising to £700 for a family with children (the equivalent of around £60 a month).

Costs are also incurred by local authorities as they have to employ people to remove household waste and take it to landfill sites.

Ways to reduce food wastage

There are many recommendations being provided on how to reduce the amount of food being wasted; some of these include:

- planning meals ahead so that you only buy what you need
- using up foods that you have at home stored in cupboards or in the freezer first
- cooking only the amount you need instead of cooking a bit extra

- using a portion size measuring device to ensure you only cook the amount of portions you need.
- using leftover foods to create different meals the next day
- freezing any unused portions of meals that are suitable for freezing to save time another day and to save wasting the one portion of stew that was left over.

> ### KEY POINTS: FOOD WASTE
> - Food waste is a major concern in the UK.
> - The cost of wasting food can be up to £700 a year for a family.
> - A wide variety of foods is wasted including fruit and vegetables, salad and bread products.
> - Foods that are thrown away end up in landfill sites which can affect the environment.
> - Following some simple rules such as planning meals and using portion size measuring devices can help to reduce the amount of food wasted.

Exam practice questions

1   Name **three** examples of foods that are thrown away as waste.   [3 marks]

2   Give **two** reasons why we waste so much food.   [2 marks]

3   Suggest **two** ways to prevent food wastage.   [2 marks]

SECTION 5

Where food comes from

This section contains the following chapters:

CHAPTER 8
Food provenance

Learning objectives

In this chapter you will learn:

■ where foods come from and how they are grown, reared and caught
■ the meaning of food miles, the impact they have on the carbon footprint, and about buying foods locally
■ the value of packaging and the impact it has on the environment
■ the sustainability of food and how to be sustainable
■ the impact of food waste on the environment
■ the differences between local and global markets and communities
■ the meaning of food poverty and the impact it has
■ food security and the access to safe, sufficient food.

Food origins: where and how foods are grown, reared, or caught

How foods are grown

There are many different foods grown in the UK. Some of the main crops grown in the UK include:

- wheat
- barley
- oats
- potatoes
- sugar beet
- vegetables (such as carrots, cabbage, runner beans etc.)

- oilseed rape
- fruits (such as apples, pears, plums, damsons; and soft fruits such as strawberries, raspberries, blackberries and blackcurrants).

Figure 8.1 Wheat growing in a field

The UK does not produce all the food that is required to feed everybody due to the population size and the climate, so some food is **imported** from other countries. An example of this would be some of the fresh fruits and vegetables (for example, exotic fruits) that are available in supermarkets.

There is a range of methods used for growing. How the foods are grown will depend upon the type of foods being grown.

Farming is an example of how foods are grown in the UK. The main two forms of farming are dairy farming and general farming for crops.

How are crops grown?

When growing crops such as wheat and barley, a number of points must be considered by the farmer. Some of these include:

- general weather conditions for the area
- which crops will grow well together
- the nutrients the crops will need
- the resources the farmer has, including natural resources (land, air, sunlight), tools and equipment, as well as manpower.

The process of growing crops includes stages such as:

- preparing the soil
- sowing seeds or plant seedlings
- watering
- controlling crop pests, fertilising, as well as weed control
- harvesting.

Figure 8.2 Crop production

The eastern and southern areas of England tend to be the best areas for growing cereal crops due to the fields being more flat and open. Some areas in Wales and Scotland focus more on sheep and dairy farming due to the land being hillier and therefore unsuitable for producing crops.

Growing vegetables

Many of the staple vegetables we eat are grown in fields on large farms. Some farms specialise in producing large quantities of one vegetable (for example, carrots) but provide a number of different varieties, providing supplies for many supermarkets and shops.

Growing soft fruits

Soft fruits are popular foods during the summer months when they are in season. Many soft fruits are grown by privately owned, family-run farms.

Figure 8.3 Carrots growing in field

In the UK, most soft fruits are grown in a **polytunnel**. A polytunnel is a large polythene tunnel that is designed to protect the fruit from poor conditions such as the weather. The use of a polytunnel will also guarantee a better quality product and reduce the need for pesticides. Using this method of growing the fruits and other foods such as lettuces, cucumbers and peppers means the growing time for crops is extended.

Growing hard and stone fruits

Many hard fruits and stoned fruits are grown in the UK. They can be produced on a small scale, large scale or be commercially produced. Some counties in the UK such as Kent, Worcestershire and Herefordshire are popular areas for growing fruit trees due to the correct combination of soil and sunlight. Many household gardens grow fruit trees successfully.

Figure 8.4 A polytunnel used for growing strawberries

Figure 8.5 Apples growing in an orchard

Hard fruits such as apples can come in many different varieties and colours. Many apples are grown in orchards that are designed to make the most of the space available and maximise production.

Activity

1 Carry out some research on the fruits being sold in your local shop or supermarket. Find out which ones are grown in this country and which ones have been imported.
2 Draw up a chart to present your findings.
3 Discuss your findings and make comparisons in relation to time of year, season and climate.

Intensive farming

Intensive farming is a method used by many farmers to increase food production. In intensive farming, fertilisers and pesticides are used to grow **high-yield** crops such as wheat, and animals are kept indoors, for example battery farming.

Figure 8.6 An organic farm

Organic farming

Organic farming is when food is produced and grown as naturally as it can be under strict standards. There is a strong emphasis on farmers ensuring the protection of wildlife and the environment.

Unlike intensive farming methods, in organic farming artificial chemical fertilisers are not used; instead organic farmers will use organic matter to help develop a healthier, fertile soil, and encourage wildlife to help to control pests and disease.

Crop rotation is encouraged; this means a farmer will plant a crop in a field one year and the next year animals are allowed to graze on the same field, adding manure to the soil and improving the fertility. The next year the field could be left empty (fallow) allowing the soil to recover.

KEY WORDS

Imported: to bring foods into a country from abroad for sale.

Polytunnel: large plastic tunnel/ frame to cover foods when they are growing.

High-yield: grow in abundance; grown to produce large results.

How foods are reared

Many animals are reared on a large scale in factory farms to provide food to be sold in shops and supermarkets. Some animals are reared on a smaller scale, in family-owned farms or one-animal specialist farms.

It is estimated that each year approximately one million animals are killed to provide food.

Some examples of animals reared for food include:

- cows and calves
- pigs
- sheep and lambs
- chickens
- turkeys
- geese
- ducks.

Figure 8.7 Chickens reared in a factory farm

Factory farms

The aim of a **factory farm** is to maximise the number of animals that can be reared. Factory farming focuses on profit and efficiency rather than the welfare of the animals. The conditions the animals are reared in are often described as basic or poor. Animals may have minimal space to move in, are not allowed to roam around, and are fattened up quickly; this can be through the use of drugs.

Many animals are reared in factory-style farms and kept to produce foods. For example, dairy cows will produce milk to be sold in supermarkets as well as producing milk to be used in the manufacturing of other dairy foods such as cheese. The cows live in large sheds, which can have limited or sometimes no access to sunshine or to pasture. Cattle cubicles, which have limited space are used for the milking process.

Organic farms

Animals reared on an **organic farm** will have very different living conditions to factory-farmed animals; the welfare of the animal is always put first and there are organic standards that have to be met. Some of these include:

- Animals are to be free-range and must have access to fields (if weather conditions are suitable).
- The living conditions have to meet high welfare standards and the animals must have a certain amount of space.
- The diet has to be as natural as possible.
- The animals must only be given drugs to treat an illness.
- The animals cannot be given hormones which make them grow more quickly.

Figure 8.8 Pigs outside on an organic farm

KEY WORDS

Factory farms: these farms maximise the number of animals that can be reared. Factory farming focuses on profit and efficiency rather than the welfare of the animals.

Organic farms: these farms put the welfare of the animal first and there are certain standards that have to be met.

How foods are caught

Many different types of fish are caught for food. Some of these include:

- oily fish – mackerel, herring, salmon, trout, tuna
- white fish – cod, plaice, haddock, sole, halibut, sea bass
- shellfish – lobster, crab, prawns, mussels, scallops.

There are many different methods used to catch fish. The method used will depend on:

- the type (species) of fish being caught
- the area where the fishing is taking place
- the scale of the fishing – small scale or large scale
- the technology available to the fisherman.

Some fishing methods, which do not catch a large amount of fish, can have a low impact on the environment compared to commercial fishing which is on a large scale and aims to catch enormous amounts of fish. Large-scale fishing can cause problems in the marine environment.

Most of the fish we eat are caught by **trawling**. This uses a net or nets to catch the fish. There are many different methods of trawling, which use nets in different ways. Some examples include pelagic trawling, otter trawling, beam trawling and pair trawling.

Other methods used to catch fish include:

- **purse-seining** – drawing a huge net around a school of fish, for example tuna, herring, mackerel
- **dredging** – towing metal cages across shellfish beds.

Stretch & challenge

Do some research into the following different trawling methods and explain how they are used to catch fish:

- pelagic trawling
- bream trawling
- otter trawling
- pair trawling.

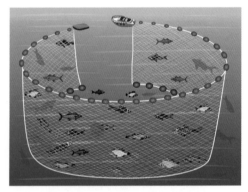

Figure 8.9 Purse-seining

Figure 8.10 Otter trawling

Farmed fish

There are many species of fish that are raised in fish farms, due to the reduced amount of wild fish available. These include:

- salmon
- trout
- cod
- sea bass.

Figure 8.11 A fish farm

In this method, fish are reared in large numbers in rivers or in tanks, enclosures, lakes, or at sea in sea cages, which can be overcrowded due to the amount of fish they contain.

Organic fish farms will apply the same standards for **farmed fish** as they do to animals raised on land.

Lobster and crab pots

Lobster and crab pots are a method used to catch shellfish. The pots can be made out of wood, rope or metal and are placed on the seabed using ropes. Dead fish is used as bait, which encourages crabs and lobsters to enter the pot. Once in, they cannot get out. Fishermen will use a buoy to mark the location of the pot and will collect their catch after one or two days. This method of fishing is regarded as being more 'friendly', as any other fish caught in the pots can be put back into the sea.

Figure 8.12 Lobster pots

KEY POINTS: HOW FOODS ARE GROWN, REARED AND CAUGHT

- Wheat, barley, oats, potatoes, fruits and vegetables are some of the main crops grown in the UK.
- Farmers must consider a number of key points before growing crops including the weather conditions and nutrients required.
- Many soft fruits are grown in a polytunnel.
- In the UK, many hard and stone fruits are grown, including apples, cherries and plums.
- Intensive farming is used to produce foods on a large scale.
- Organic farming grows and produces foods as naturally as possible.
- The two main methods of farming animals are factory and organic.
- The living conditions and environments are very different in factory and organic farms.
- Oily fish, shellfish and white fish are the main types of fish caught for food.
- There are many different methods of sea fishing used to catch fish. Most are caught by some form of trawling.
- Fish farms are used to raise fish due to decreasing availability of wild fish.
- Lobster and crab pots are methods used to catch shellfish.

Exam practice questions

1. List **three** of the main crops grown in the UK. [3 marks]
2. Explain the difference between intensive farming and organic farming. [4 marks]
3. Describe why some customers will choose organic produce when buying food. [6 marks]
4. Name **three** different animals reared for food. [3 marks]
5. Explain why many people could be against factory farming for animals. [2 marks]
6. State **two** organic standards that must be met in organic farming. [2 marks]
7. Name **two** oily fish and **two** shellfish that are caught for food. [4 marks]
8. Describe **two** different methods used to catch fish. [2 marks]
9. Explain how farmed fish differ to fish caught at sea. [2 marks]

Food miles

Food miles is a term used to describe the distance that food is transported from the field or the food producer to the plate or consumer. This also includes the miles the consumer has travelled from home to the shop to buy the food.

Many of the foods available to buy in shops are from farms and factories that are in towns or rural areas that are hundreds of miles away from the shop. This could be due to a number of reasons. For example, the crops required to produce the foods could have been grown in that particular area, or the food manufacturer may have its main processing plant in that area. These foods or food products are then transported all over the country to various shops, so that consumers can buy them.

Some fruits and vegetables available in our shops and supermarkets come from other countries. This is because these foods are not in season or cannot be grown in the UK, so would not be available for us to buy if we did not import them.

Importing foods from other countries means that we can buy foods such as strawberries and asparagus when we want to at many other times of the year rather than just when they are in season.

When foods travel or are imported they could be transported by:
- boats (sea)
- aeroplanes (air)
- lorries (road)
- cars (road).

Carbon footprint

The meaning of the term **carbon footprint** of a product is the amount of carbon emissions that have been produced during the growing, processing and distribution (transporting) of a food product. In basic terms, this means all of the processes that must be gone through to create a final product.

The carbon footprint includes the estimated amount of carbon dioxide (CO_2) given out as vehicles travel.

Importing foods can have a huge impact on the carbon footprint, and transporting foods by air has more impact than transporting foods by sea or road – although these methods are still considered to be an environmental problem.

When foods are being transported in lorries by road, the lorries burn fuel, producing large amounts of air pollution.

When foods are being transported by road, sea or air, fossil fuels are burned. This affects the environment because when fossil fuels are burnt they release carbon dioxide gas emissions, which have a big impact on **global warming**. Global warming is described as the gradual heating of the Earth's surface, oceans and atmosphere.

Reducing transportation emissions is one of the most important steps that needs to be taken to fight global warming. Another step is to plant more trees, as these will absorb the carbon dioxide. Buying local produce can also help the environment.

Figure 8.13 Food being transported by road

Buying foods locally

Buying foods locally means purchasing items near your home – this could be from the shop around the corner, from the greengrocers in the local town, or from the nearby farm that sells meat. This means that the foods were grown, assembled or manufactured nearby.

By purchasing foods that are produced and grown locally you:

- support local growers
- can help the local community earn money
- can help create local jobs
- will be able to buy the foods for less money
- will be able to buy fresher, better quality produce
- will be able to buy foods that have a higher nutritional value
- will be able to buy seasonal goods as they are at their peak
- can ask the local growers questions about the food directly
- can buy safer food as the food chain is minimal compared to imported goods.

Figure 8.14 A food stall at a local farmers' market

Practical activity

1 Make a tropical fruit salad using a selection of fruits.
2 Demonstrate accurate knife skills and the correct hold to cut the fruits into even-sized shapes and slices.
3 Present your fruit salad, making it look as attractive as you can.

Activity

1 Visit your local supermarket or shop and look at the selection of tropical fruits that are being sold. Make a note of all the different countries the fruits have come from. Record your results in a table showing the nearest countries to the UK first and ending with the countries that are the furthest away.
2 Calculate the total number of air miles that the fruit from furthest away has travelled.
3 Explain how all the travelling the fruit has done would have harmed the environment.

KEY WORDS

Food miles: the distance the food has travelled from field or production to the consumer's plate.

Carbon footprint: the amount of carbon emissions that have been produced during the growing, processing and distribution (transporting) of a food product.

Global warming: the gradual heating of the Earth's surface, oceans and atmosphere.

KEY POINTS: FOOD MILES, CARBON FOOTPRINT AND BUYING FOODS LOCALLY

- Food miles is the distance the food has travelled from field or production to the consumer's plate.
- Foods can have a large amount of food miles if they are brought to this country from countries abroad.
- Food miles can refer to foods that have travelled by car, boat, plane or lorry.
- Carbon footprint refers to the amount of carbon emissions that have been produced during the growing, processing and distribution (transporting) of a food product.
- The carbon footprint is measured by the amount of carbon dioxide given out, and this has an impact on global warming.
- Global warming is the gradual heating of the Earth's surface, oceans and atmosphere.

Exam practice questions

1 Explain the benefits of buying locally produced fresh vegetables instead of imported fresh vegetables. [6 marks]

Packaging

Most of the foods we buy in shops are sold in some form of packaging. Some foods that are not sold in packaging include loaves of unsliced bread, bread rolls, loose fruits and vegetables. When consumers buy these products they often put them into plastic or paper bags supplied by the shop or supermarket, so they do still have some form of packaging.

There are many different types of packaging used for foods and food products; the type of packaging will often depend upon the foods or food products being sold.

Why are foods packaged?

Manufacturers will package food to:

- keep the food safe and hygienic
- protect the food from damage or contamination
- preserve the food and extend its shelf life
- prevent **tampering** with the food
- attract customers to buy it
- make the food more presentable for selling
- provide information about the food
- make handling, transportation and storing easier.

Figure 8.15 Information provided on the packaging of a food product

The value of packaging for manufactured foods and food products

Food packaging is used to market a product. It is a method used to encourage consumers to buy a particular product. The outside packaging will provide consumers with essential information such as:

- how to store the food product and the advised length of storage
- which ingredients the product contains
- how to cook it or heat it up
- what the product looks like
- the name of the product
- who manufactured the product.

Types of packaging materials used

Type of packaging material	Examples of food
Card/paper	Breakfast cereals, bags of flour
Glass	Sauces, pickled foods
Plastic	Yoghurts, fruits and vegetables
Metal, foil and cans	Canned vegetables, soup, fizzy drinks
Ovenable paperboard	Chilled and frozen ready meals, some pizzas

Table 8.1 Examples of which foods use certain packaging materials

What impact does packaging have on the environment?

Food packaging can affect the environment in a number of ways:

- Producing the packaging can use up natural resources, for example trees and oil and huge amounts of energy.
- It can cause pollution – this could be air, land or sea pollution.
- It cannot always be recycled due to the mixed materials that have been used in the packaging. For example, some fruit juice cartons are made up of several different layers **laminated** together (e.g. card, plastic and foil).
- If it is not **biodegradable** it can take many years to breakdown. For plastics made from oil this could be hundreds of years.
- Unrecyclable packaging is disposed of in landfill sites which take up space and cause pollution.

KEY WORDS

Tampering: to interfere with the food to cause damage.

Laminated: many different layers of materials bonded together.

Biodegradable: a product or material that can dissolve easily in the environment without destroying nature.

KEY POINTS: PACKAGING

- Foods are packaged for a number of reasons including protection, preservation and presentation.
- Packaging can provide consumers with essential product information.
- There are many different types of packaging materials used including glass, plastic, paper and card.
- Packaging can be quite environmentally damaging due to the materials and processes used.

Exam practice questions

1   State **three** reasons for packaging foods.   [3 marks]
2   Give **two** reasons why a consumer could find the packaging of a food product useful.   [2 marks]

3   Explain why packaging is considered to be harmful to the environment.   [4 marks]

Sustainability of food

Sustainability is about meeting a variety of needs for us and others, and future generations to come. It is about meeting the needs of the country's population, as well as globally, in relation to food availability and the growing of food using sustainable systems. It is about considering where our food comes from and the process from field to plate, the necessity of avoiding food wastage and reducing the impact on the environment, and the importance of buying local and seasonal food, taking into consideration the farming and growing methods used to provide food.

The impact of food waste on the environment

As a nation we throw away billions of pounds worth of unused food every year. If the food that we waste goes to landfill sites, it will go through the process of decay and eventually rot. When food rots it produces methane; this is a powerful greenhouse gas.

Reducing food waste

We are therefore advised to try to reduce our food wastage. There are many ways to try to reduce the amount of food that we waste; these are discussed in Chapter 7 Food spoilage.

Recycling

Some of the leftover food or peelings from preparing foods that we throw away could be put to use by recycling them in a compost heap or food recycling bin. This would be suitable for leftovers such as:

- fruit and vegetable peelings
- cooked pasta or rice
- stale bread.

This will rot down to produce fertiliser that can then be used in the garden by adding to the soil when planting seeds.

Reusing food

Reusing leftover food is a positive step that can be taken to reduce the amount of food we waste, as well as reducing the impact it has on the environment. By reusing leftover foods, the amount of money that is spent on food would be reduced, and the nutritious ingredients could be used to produce more home-cooked meals. Ways to reuse leftover foods could be:

- Leftover vegetables from a roast dinner could be used to make a soup or bubble and squeak.
- Leftover pasta could be used to make a pasta salad.
- Stale bread could be used to make bread and butter pudding, or breadcrumbs to go on top of pasta dishes to make a crunchy top.
- Leftover mashed potato could be used on top of some leftover vegetables that have been mixed with a cheese sauce to make a potato-topped pie.
- Leftover sponge cake could be used in the making of a trifle.

Local and global markets and communities

Local markets

As people are becoming more aware of the impact on the environment when foods are imported, the demand for local food markets is increasing. Farmers' markets have become very popular; they sell food that has been grown and sourced from local farms around the area in which the market is being held.

The benefits of using a local or farmers' market include:
- consumers get fresh, healthy produce, usually at competitive prices
- the markets offer increased choice of foods that are very fresh and normally affordable
- they strengthen the local community
- they can help to reduce the carbon footprint due to limited travelling and pollution.

Global markets and communities

As a nation we have grown used to having a plentiful supply of food that is affordable. Some of this food supply is sourced from the UK, and some of it is sourced from outside the UK. When consumers go into a supermarket to buy food products, the range of foods and choices available to them has increased considerably compared to 40 years ago.

The world population is increasing, and therefore the amount of food available is decreasing. There is a global demand to feed this growing population. With people's incomes increasing, the population is also richer and more money is available to spend on food. There is a lot more competition for food, and as this increases, it could limit our access to the same types of foods that we are used to.

The food supply chain has changed over the years and can now involve many different countries. For example, rice grown in Thailand may be packaged in India and sold in a supermarket in the UK. Food is transported to where the demand is.

Consumers are being encouraged to use local ingredients and local suppliers because of the increased emphasis on food sustainability.

Community farming

Community farming is becoming more popular in many areas of the UK. Land is often rented from a local farmer. The farm is community owned and run and can have a number of people investing in the scheme so that is it feasible. Community farms can promote social gatherings, bringing together many different people from the local area.

The food grown is organic and the farms aim to provide good-tasting, fresh organic produce direct to the consumers within the local community. The delivering of organic vegetable boxes is also a method of selling the local produce. Community farms try to develop people's understanding of where food comes from; they look to reconnect with the land on which their food is grown, as well as learn more about sustainable farming.

The effect of food poverty

Food poverty means that an individual person or household is not able to obtain healthy, nutritious food. The type of food that people in this situation tend to consume is often low-cost food that is high in fat and sugar, which can then lead to diet-related issues such as heart disease and diabetes. Sometimes people find themselves having to go without food, or missing out on meals due to not having sufficient money available to buy food.

According to the Joseph Rowntree Foundation report *Monitoring Poverty and Social Exclusion*, in 2010 there were 5.8 million people living in food poverty in the UK.

There are many different charities in the UK that provide information about the rise in food poverty. Oxfam and FareShare are just two examples. FareShare, for example, highlights that 3.9 million tonnes of food is wasted every year by the food and drink industry, with 10 per cent of the food being surplus and fit for consumption. The charity aims to make use of this surplus by linking with businesses in the industry and distributing the leftover food to people who are considered in need due to food poverty.

Food poverty is on the increase for a number of reasons:
- Low incomes and shrinking incomes mean people cannot afford to buy food as well as pay other household expenses.
- Rising fuel bills mean families are struggling to heat their homes and feed the family.
- Rising food prices mean families have to buy less food due to higher costs.
- Cuts and changes to the benefits system.
- People having to pay back debts or loans, leaving them with no money for food.
- People being out of work and not being able to provide money to buy for the family.

Problems for children

Many children will not be eating the right foods to provide them with the essential nutrients they need; this can lead to deficiencies and poor dietary patterns. During term times, many children will have a cooked meal provided for them by the school. In the holidays this can then mean that they go without and experience hunger. Many parents will skip meals to feed their children.

Food banks

The use of food banks is increasing, as many families rely on this 'helping hand' to be able to obtain food. The food can be donated by businesses, churches, schools, and individual people who will buy an extra item when they shop. The foods are non-perishable; this means they have a long shelf life and do not require refrigeration. They are handed out when the people in need hand over a voucher.

Figure 8.16 A food bank

Activity

1 Carry out some research in your local area on food banks. Find out if there are any, and if so, how many are there and where the nearest one is.

2 Look in your local shop and supermarket to see if they collect donated foods from customers to give to the food banks.

   a) Find out the different types of foods that are donated.

   b) Suggest some recipe ideas that could be made using the donated foods.

KEY POINTS: SUSTAINABILITY OF FOOD

- We generate a lot of food waste and should aim to recycle or reuse to reduce the impact of food waste on the environment.
- Local and farmers' markets are becoming more popular. These types of markets do not sell produce that has experienced excessive transportation.
- The population of the UK and the world is increasing, which means food supplies are decreasing.
- The food supply chain can now involve many different countries, not just one single country.
- Food poverty is increasing for a number of reasons such as low incomes, rising fuel prices and rising food prices.
- The amount of food wastage classed as 'fit for consumption' is increasing and some charities are aiming to make use of this to help people in need.
- Food banks help many different people with food supplies that are in date and non-perishable.

Exam practice questions

1 Name **two** dishes that could be made using leftover roast chicken. [2 marks]

2 Suggest **two** ways of using leftover cooked rice. [2 marks]

3 Explain the safety issues regarding the reheating of cooked rice. [2 marks]

4 List **three** benefits of shopping at a local market. [3 marks]

5 Explain why the demand for food has grown over recent years. [2 marks]

6 Describe the benefits of community farming. [4 marks]

7 State **three** reasons why food poverty has increased. [3 marks]

Food security

The World Food Summit of 1996 defined **food security** as existing 'when all people at all times have access to sufficient, safe, nutritious food to maintain a healthy and active life'.

The concept of food security is defined by the World Health Organisation (WHO) as including both physical and economic access to food that meets people's dietary needs as well as their food preferences.

Food security is built on three main features that must be consistent all year round:
- food availability
- food access
- food use.

Food availability

Food availability is related to sufficient amounts of food being available regularly, as well as the source of the food being trustworthy. Food availability considers the amount of food that a country can produce for itself, the amount of food that needs to be imported and the amount of food that is exported.

Food access

Food access is related to the cost of food and the growth and distribution of food.

Food has to be available for people to buy at an affordable price. Families on reduced incomes or low incomes may not be able to afford good quality food.

There has to be sufficient land available for growing crops or rearing animals to provide food to meet the different demands. Problems that can occur due to weather – for example, if there is too much or too little rain – can affect the growing of crops.

Once food is produced, considerations include how it will get to where it is needed, whether there is suitable transportation, and whether the food will be shared out appropriately.

Food use

Food use is related to how well people manage the food. Food use is determined by how well informed people are about the nutritional value of the foods, and whether they put this knowledge into practice when preparing and making different meals. Food should be used to prevent illness and disease and to maintain good health.

Key factors that can have an impact on food security

Food security is not just one individual element; it is many different elements that link together and affect many different countries.

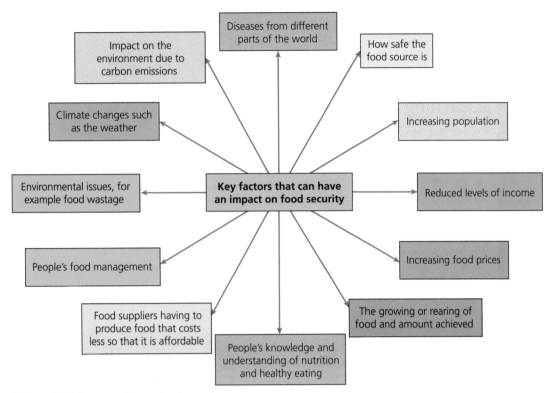

Figure 8.17 Factors affecting food security

 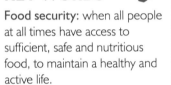
KEY POINTS: FOOD SECURITY

- Food security is a global issue concerning many different countries.
- Food security relates to three main features: food availability, food access and food use.
- Key factors that have an impact on food security can include climate, increased populations, reduced incomes and diseases around the world.

Exam practice questions

1 Explain the meaning of food availability. [4 marks]
2 State **two** factors that are related to food use. [2 marks]

3 Name **three** factors that can have an impact on food security. [3 marks]

CHAPTER 9
Food manufacturing

Learning objectives

In this chapter you will learn about:

■ the features, characteristics, eating patterns, recipes and meal structures of different cuisines, including British cuisine and a range of international cuisines
■ primary stages of processing and production
■ secondary stages of processing and production and how primary products are changed into other types of products
■ how processing affects the sensory properties of ingredients
■ technological developments that support food processing and production
■ the positive and negative effects that food modification may have on health
■ the use of additives.

Culinary traditions

When using the term **cuisine**, we are referring to the traditions of a country in relation to types of foods, meals, recipes and the different styles of cooking associated with the country.

British cuisine

Traditional British food could be described as hearty, fulfilling, substantial, and at times simple in its flavours, as well as wholesome to eat.

Traditional dishes

Traditional dishes are regarded as typical dishes or meals associated with a cuisine. Some of them could be from a particular area of the country; some could be made using a particular method, or include a key ingredient associated with the country, for example potatoes, which are a **staple** British food. Other examples of typical British staple foods could include bread, flour and milk.

Examples of traditional British dishes include:
- steak and kidney pie
- fish and chips
- bangers and mash
- full English breakfast
- roast beef and Yorkshire Pudding
- beef stew and dumplings
- Cornish Pasty
- steamed sponge pudding and custard
- treacle tart.

Figure 9.1 A traditional roast dinner

Modern British cuisine

British cuisine has changed over the years and it could now be described as having a more modern approach to it, with traditional dishes being reinvented through the style of cooking used. An example of this would be the cooking of the meat for a traditional roast dinner. Traditionally, this would have been cooked in the oven, sometimes for a long period of time, which could result in a dry texture; or if cooked for a short period of time, the meat could be undercooked. A more modern approach to cooking meat would be to use **sous vide**. In this method, food is vacuum-sealed in a pouch and then cooked slowly at low temperatures. The food becomes tender without losing its original colour, nutrients and texture.

The style of presenting foods has also changed as the style of British cuisine has developed. Dishes are being reintroduced with a modern twist to their appearance and style. The roast chicken meal is an example of the modern approach.

Figure 9.2 Comparison of cooking methods

Meal structures

In Britain the eating pattern is generally to eat three meals a day: breakfast, lunch and dinner; although for some people, high tea is also consumed in the middle of the afternoon and then dinner or the evening meal is eaten later.

Most people from other countries often imagine a typical British breakfast to be eggs, bacon, sausages, fried bread, mushrooms and baked beans, all served with a cup of tea or coffee. However, it is more common to describe a British breakfast as a bowl of cereal, a slice of toast, orange juice and a cup of tea or coffee.

A traditional dinner or evening meal can consist of meat and 'two veg'. Hot brown gravy, traditionally made from the juices of the roast meat is often served over the meat. One of the vegetables eaten is almost always potatoes, as potatoes are a staple food of Britain.

Figure 9.3 Roast chicken

International cuisines

There are many different international styles of foods, dishes and meals, which are available in supermarkets and restaurants for the consumer to experience. Examples of some of the cuisines you may find in your high street or area are shown in Table 9.6. The style of dish and types of ingredients used can vary considerably according to region and or state.

Type of cuisine	Example dishes	Key ingredients	Cooking methods
American	Burgers, barbequed ribs, fries, potato salad, corn on the cob, cornbread, hot dogs, fried chicken, clam chowder, meat loaf	Beef, pork ribs, potatoes, sour cream, cheese, baked beans, potatoes, herbs e.g. rosemary, pickles	Hotplate frying, deep-frying, marinating and cooking over coals like a barbeque, grilling, stewing
Chinese	Chicken chow mein, sweet and sour pork, spring rolls, Peking roasted duck, wontons, dumplings	Rice, noodles, vegetables, eggs, fish, tofu, meat (pork is the most common), soy sauce, scallions, bean sprouts, cabbage, ginger root, aniseed, cinnamon	Stir-frying using a wok, steaming, blanching, deep-frying
French	Moules marinière (mussels in a white wine broth with shallots and parsley) Sole meunière (sole fried in butter with added lemon juice) Cheese soufflé, steak tartare, soupe à l'oignon, boeuf bourguignon, tarte tatin	Seafood, meat, different cheeses, vegetables including onions, bread (baguette), herb mixtures to include thyme, sage, rosemary, basil, savory, fennel, marjoram, tarragon, oregano and bay leaf	Braising, flambéing, (the addition of alcohol to the food in the pan to produce flames that are immediately burnt away to leave the aroma and flavour), grilling, poaching, sautéing
Italian	Minestrone soup, pizza, lasagne, antipasto (small bites before the meal), olives, meats, breads, wild mushroom risotto, ravioli, panna cotta, biscotti	olive oil, pasta, tomatoes, garlic, rice, corn (for polenta), cheeses for example; ricotta, parmesan, mascarpone and mozzarella, seafood, meats e.g. salami, prosciutto, pepperoni, herbs including basil, pine nuts	Pan-roasting, oven-roast, sautéing, braising, broiling (cooking under high heat like a grill) grilling, pan-searing, simmering and baking
Indian	Biryani (mixed rice dish), chapathi, naan bread, dal (assorted types of lentils), dum aloo (potatoes cooked in curry) samosa, poppadoms Different types of curries or dishes named after methods used to make them: kormas, jalfrezi, biryani	Herbs and spices e.g. coriander, whole or powdered chilli pepper, cumin, turmeric, ginger, black mustard seeds, garam masala, also almonds, coconut, mung beans, lentils, chickpeas, rice, ghee, as well as other main ingredients like meat, for example mutton, lamb, chicken, beef	Many dishes are prepared by shallow-frying ingredients like onions and spices then braising the meat for a long period of time. Many meats are marinated to get the flavour in. Many foods are steamed or smoked. A karai is used (similar to an Asian wok); a tava is used (cast iron flat griddle); Bhuno is a combination of sautéing, stir-frying and stewing; the tandoor is used (a circular clay oven fired by wood or charcoal)
Mexican	Chilli con carne, tacos, guacamole, enchiladas, elote (Mexican corn on the cob served in cups with salt, chilli powder, lime, butter, cheese mayonnaise and sour cream), chilaquiles (a breakfast dish: fried corn tortillas cut into quarters and topped with green or red salsa, scrambled eggs can also be added on top), churros	Spices such as oregano, cumin, chilli powder, cocoa, avocado, beans (e.g. pinto), rice, herbs such as coriander, tortillas (corn and white flour) and chipotles (smoked, dried jalapenos), traditional meats (e.g. beef and pork), seafood, seasonal vegetables	Can include stir-frying, deep-frying, baking, grilling, boiling, simmering, marinating. Cast iron pans are used to make tortillas, mortar and pestle are used to grind and mash

Table 9.1 Types of international cuisine

Meal structures

Many European countries have the same eating pattern as British cuisine: breakfast, lunch and an evening meal. Often lunch can be the most important meal and in some countries (for example Italy) many places of work like shops or banks will close early afternoon to allow workers to go home or dine out for lunch.

A typical Italian lunch will consist of a first course – this could be pasta or a rice-based dish. The second course would be meat, fish or vegetables. Fruit is often eaten to end the meal.

If the meal is related to festivities or a celebratory occasion then the number of courses can increase and the meal could involve:

- An aperitivo – to begin the meal. This course may consist of a glass of a bubbly beverage like prosecco or a small dish of olives.
- Antipasti – this course is considered to be the starter and could be a cold traditional meat platter including salami and prosciutto. It can be served with cheeses and bread.
- Primi – this is the first course to contain hot food and does not always contain meat; risotto would be an example of a dish served.
- Secondi – this course will offer you different meat or seafood options. Vegetable-based side dishes will be served as well.
- Formaggi e frutta – seasonal fruits and regional or locally produced cheese will be served in this course.
- Dolce – this would be the dessert course; an example would be tiramisu.

Within different cultures, for example Chinese and Indian, the meal structure can be quite different compared to the European style. A large number of dishes can be put onto the table at the same time for people to select from, offering them a variety of flavours and foods as shown in the image below.

Figure 9.4 **A selection of Indian dishes**

1 Collect a menu from an international restaurant in your area. Read through the menu and plan what you would eat for a three course meal if you were having a meal in the restaurant. Write down your choices.

2 Add up the cost of the meal and record the total cost.

3 Research on the internet or in books and find a recipe for the main course you have chosen.

4 Calculate the cost of making the dish using a supermarket site.

5 Explain your findings – did it cost more to make the dish yourself, or was it less expensive to make it yourself? Why do restaurants prices cost more than it costs to produce a dish at home?

1 Choose a dish from the menu you have collected that you would like to make. Research to find a recipe you could use.

2 Cook the dish you have chosen and present it to your family. Ask them to taste test it and give you some suggestions on how the dish could be improved and/or adapted.

3 Consider if the dish needs to be served with any accompaniments. You could further develop your practical skills by making some accompaniments to be served with it next time you make it.

KEY WORDS

Cuisine: a style characteristic of a particular country or region with which specific ingredients, cooking methods, equipment, presentation and serving styles are associated.

Sous vide: a cooking method in which food is cooked slowly at low temperatures in a vacuum-sealed pouch.

KEY POINTS: CULINARY TRADITIONS

- **Cuisine** refers to the traditions of a country in relation to types of foods, meals, recipes and the different styles of cooking associated with the country.
- Traditional British food includes hearty dishes such as fish and chips, roast beef and Yorkshire pudding and full English breakfast.
- British eating patterns tend to include three meals a day: breakfast, lunch and dinner.
- Modern variations on British cuisine include slower cooking at lower temperatures (e.g. sous vide) and traditional British dishes presented with a modern twist.
- European eating habits can be very similar to British eating habits, with three meals a day.
- Other international cuisines include Chinese, Indian and American, and the style of dish and types of ingredients used can vary considerably according to region or state.

Exam practice questions

1 British cuisine has changed over the last couple of decades. Explain how this style of cuisine has changed to have a more modern approach.

[4 marks]

Primary stages of processing and production

One of the first or primary stages of processing foods is to grow or rear a food so that it can be changed or transformed into a suitable state to either be eaten or used in the production of other products.

A **primary food** is not edible in its original state; it has to be changed or have some form of preparation before it can be eaten. An example would be a raw potato. Sometimes, the **primary processing** can be quite basic, for example peeling vegetables or washing salad leaves.

Other examples of primary foods could include:

- wheat
- maize
- soya beans
- sugar beet
- milk.

Wheat is an example of a primary food. It is processed into flour, which is a secondary source of food. The stages of processing for wheat are shown in figure 9.6.

Figure 9.5 Primary processed carrots

Point of origin

The point of origin for a food item generally means where the essential character of the food is created or grown.

A potato, for example, is a staple food that can be grown in fields on farms, on allotments and in gardens. Potatoes are grown to be sold in various areas of the UK, for example Cornwall and South-West Wales. When identifying the point of origin for the potatoes, these areas would be named, as the potatoes would have been grown on farms or allotments in these areas.

Transportation of primary foods

Once grown or reared, primary foods are transported to the processing plant or factory. This could be done using one of many methods or a combination of methods. The method or methods chosen will depend upon the location of the farm or allotment on which the goods were reared or grown and where processing is taking place. Transportation could be in the form of:

- hand or mechanical harvesting equipment– goods can be picked by hand or mechanically (for example using tractors, combines, mowers, corn harvesters)
- refrigerated trucks to maintain freshness and quality and prevent foods 'going off'
- transport trailers for live cattle.

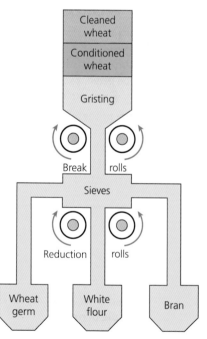

Figure 9.6 The stages of processing wheat

Cleaning and sorting of raw foods

Primary stages of processing and production also include the cleaning and sorting of the raw materials. This may be carried out to remove any unwanted debris, to clean off mud or dirt, or to remove any damaged goods. Figure 9.7 identifies stages of this process for potatoes.

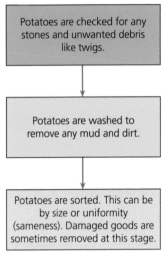

Potatoes are checked for any stones and unwanted debris like twigs.

Potatoes are washed to remove any mud and dirt.

Potatoes are sorted. This can be by size or uniformity (sameness). Damaged goods are sometimes removed at this stage.

Figure 9.7 Potatoes being sorted

KEY WORDS

Primary food: a food that is not edible in its original state; it has to be changed or have some form of preparation before it can be eaten. Examples include raw potato, wheat and maize.

Primary processing: the process of changing or transforming primary foods into a suitable state to either be eaten or used in the production of other products.

KEY POINTS: PRIMARY STAGES OF PROCESSING AND PRODUCTION

- Examples of primary foods include potatoes, wheat, maize, sugar beet and milk.
- Primary processing involves changing or transforming primary foods that are grown or reared into a suitable state to either be eaten or used in the production of other products.
- Once grown or reared, primary foods are transported from their point of origin (e.g. the field, farm or allotment) to a processing plant or factory.
- Primary stages of processing and production also include the cleaning and sorting of raw materials.

Exam practice questions

1 Explain what a primary food is and give an example of **one**. [2 marks]
2 Name **two** more primary foods. [2 marks]
3 Explain what point of origin means. [1 mark]
4 State **two** methods of transportation for primary foods. [2 marks]

Secondary stages of processing

Secondary processing is when you change or convert the primary food into an ingredient which can then be used to make a food product.

Secondary processing can provide an opportunity to create a wide variety of foods. Flour processed from wheat is a secondary product; it has been changed or converted into an ingredient that can be used to make many different food products. An example product would be bread.

To end up as a finished final food product there would be many different stages of processing involved.

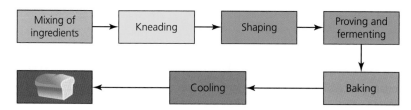

Figure 9.8 The process of making bread

Figure 9.9 shows how flour can be used as a main ingredient to make a range of different food products.

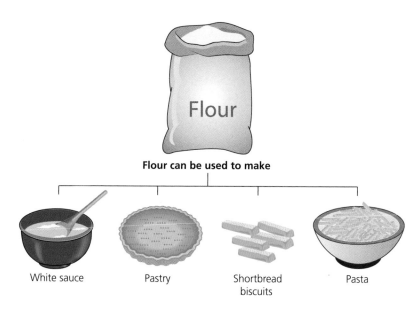

Figure 9.9 Flour can be used to make these products

There are many other examples of primary foods that are processed into secondary foods and then used to produce other products. Two examples of these are included in the table below.

Primary food	Secondary process	The product after secondary processing
Milk Taken from a cow during the milking process	**Pasteurisation** A short heat treatment to make the milk safe to drink and use	**Cheese** **Starter cultures** (these are combined single strains of bacteria) are added; these change the lactose in the milk to lactic acid which sets the protein in the milk **Yoghurt** An edible bacteria culture is added; this breaks down the lactose to form lactic acid which will thicken the yoghurt
Fruit Picked from trees/ bushes/plants, it is cleaned and could be sliced or cored	**Heating/stewing** Fruit is heated to a high temperature with sugar and water	**Jam** After heating, the fruit, sugar and pectin (a carbohydrate in the fruit) cool to form a gel which sets the jam **Jelly** After heating, the mixture is filtered/sieved to remove the pulp (flesh) so just juice remains before cooling and setting

Table 9.2 Examples of primary foods processed into secondary foods

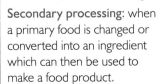

KEY WORDS

Secondary processing: when a primary food is changed or converted into an ingredient which can then be used to make a food product.

Starter culture: combined single strains of bacteria.

KEY POINTS: SECONDARY STAGES OF PROCESSING

- Secondary processing involves changing or converting primary foods into ingredients which can be used to make food products.
- Wheat can be processed into flour and used to make bread, pasta, pastry and biscuits.
- Milk can be pasteurised and have starter cultures added to it to make cheese and yoghurt.
- Fruit can be heated with sugar, water and pectin to make jam (or sieved before cooling to make jelly).

Exam practice questions

1 Explain the difference between primary and secondary processing. [2 marks]

WJEC EDUQAS GCSE Food Preparation and Nutrition

How processing affects the sensory properties of ingredients

When certain foods, for example some meats, go through the secondary processing stage they can also undergo preserving treatments, which can affect the **sensory properties** of the food. Preservation treatments are used to prevent the growth of harmful bacteria that can cause food spoilage and food poisoning, and to extend the shelf life of the food. Meats that undergo preserving treatments can have an extended shelf life of several days, weeks or even months. An example of a preserving treatment would be **curing**.

Figure 9.10 Examples of cured meats

Curing is when meat is treated with preserving chemicals, for example salt, smoke and sodium nitrite/nitrate. Examples of meats that are cured include bacon, luncheon meat and salami. Some of these processing methods can alter the sensory properties of the food, in particular the flavour and colour of the meat. For example, bacon can have a very salty taste; smoked meats can have a very strong, concentrated smoky flavour and luncheon meat can have a bright pink colour.

KEY WORDS

Sensory properties: the characteristics of a food that are linked to smell, sight, taste and feel.

Curing: when meat is treated with chemicals like sodium nitrite to preserve it for longer periods of time.

KEY POINTS: HOW PROCESSING AFFECTS SENSORY PROPERTIES OF FOOD

- When some foods go through the secondary processing stage, the processing methods can alter the sensory properties of the food. Curing, for example, affects the flavour and colour of meat.

Technological developments that support food processing and production

There are many technological developments that support food processing and production.

The increased use of computers in manufacturing

Computers are increasingly used to control production systems. This means that using computers throughout the different stages involved in the manufacturing of a food product has increased. Many manufacturers make use of computers to increase productivity and overall standards. Using computers in the manufacturing process is often referred to as **Computer-Aided Manufacturing (CAM)**.

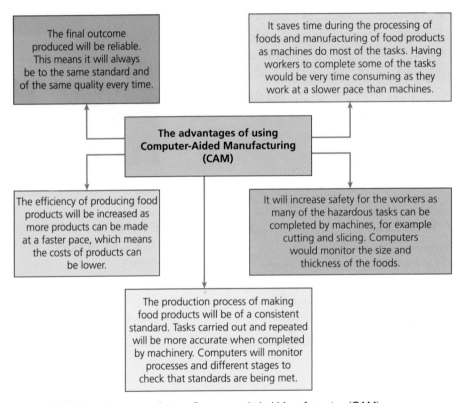

Figure 9.11 The advantages of using Computer-Aided Manufacturing (CAM)

How are computers used in the processing of food products?

Computers can be used during various stages of processing a food product. Some of the stages are identified in Figure 9.12

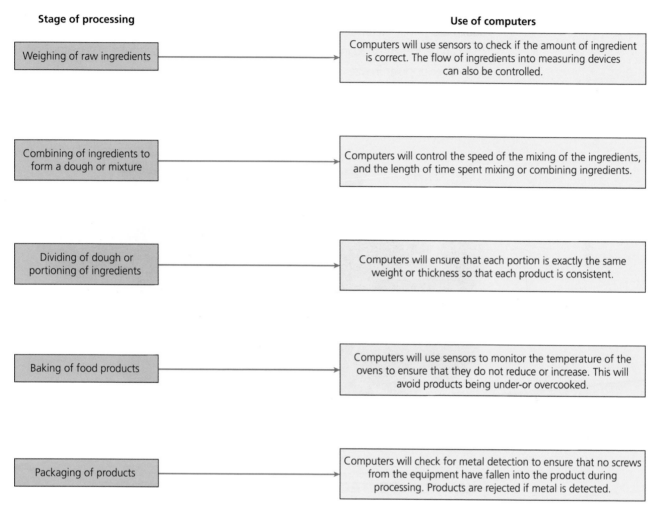

Figure 9.12 Stages in which computers are used when processing a food product

The introduction of new processes

Freeze-drying is an example of a new process. When a food has been freeze-dried it has had all of its moisture removed by a special form of drying. The food is frozen first and is then placed in a strong vacuum. The water that is present in the food will **sublimate** – that means it will turn straight from ice into vapour. Examples of foods that can be freeze-dried include some fruits like strawberries, apples and pears; coffee is another popular freeze-dried product. Freeze-drying does not have as much effect on the taste of the food as normal drying does.

Figure 9.13 Freeze-dried strawberries

> ## KEY POINTS: TECHNOLOGICAL DEVELOPMENTS THAT SUPPORT FOOD PROCESSING AND PRODUCTION
> - Computer-Aided Manufacturing (CAM) is the use of computers in the manufacturing process.
> - Advantages of using Computer-Aided Manufacturing include increased standards and productivity, increased safety of workers and time savings.
> - Freeze-drying is a new process in food manufacturing.

Exam practice questions

1 Explain in detail how using Computer-Aided Manufacturing (CAM) saves time. [2 marks]
2 Explain why computers would be used during the processing of a food product to portion ingredients or divide a mixture up. [2 marks]
3 Name **two** foods that can be freeze-dried. [2 marks]
4 Manufacturers will use computers during the processing of many different food products. Explain why a manufacturer would use sensors when baking bread products. [3 marks]

WJEC EDUQAS GCSE Food Preparation and Nutrition

The positive and negative effects of food modification on health

During food processing, manufacturers will sometimes use **additives**.

There are three different groups of additives:
- Artificial – made completely from chemicals.
- Natural – obtained from foods naturally (e.g. beetroot juice).
- Nature identical – synthetic (made chemically to be the same as a natural product).

Table 9.3 includes examples of additives and why they are used. Any additive that is used in the UK has to undergo strict testing and be approved. Numbers are allocated to additives once approved so that they can be identified. If an additive has an 'E number' this means it is accepted as being safe to use by the countries of the European Union.

Type of additive	Why used	Example foods
Preservative	• To extend the shelf life	Fruit juice, dried fruits, salad dressing
Flavour intensifier	• To improve the taste of food by adding flavour. • To restore flavours lost during processing.	Savoury snacks/savoury foods, vanilla yogurts
Stabilisers and emulsifiers	• To help foods mix together and prevent ingredients from separating out in the product when being stored. • To give foods a smooth and creamy texture. • To extend the shelf life of baked goods.	Salad dressings, low fat spreads, baked goods
Colourings	• To make foods look attractive. • To boost the colour of foods already present. • To add colour to food lost during processing.	Fizzy drinks, strawberry yoghurts, canned peas

Table 9.3 Examples of additives, why they are used and example foods they are used in

Advantages of using additives

The positive reasons for using additives during manufacturing include:
- to give consumers a wide choice of foods
- to keep foods safer for longer
- to improve a particular feature of a food, e.g. a key flavour of a food product
- to restore original features of a food product, e.g. putting colour back into food to make it look normal
- to produce a product range, e.g. different flavoured crisps
- to produce the desired effect, e.g. a creamy or smooth mouthfeel when eating a product.

Disadvantages of using additives

There are some disadvantages of using additives. These include:

- using additives to disguise lower-quality ingredients
- they can cause some people to have an allergy, e.g. skin rash
- they can cause hyperactivity.

Due to the amount of additives used in different food products and the different types that are used, there is a concern that the consumption of these on a daily basis could have an impact on health, in particular children's health. An example of this would be the linking of additives to the occurrence of hyperactive behaviour in children.

Activity

1 Look at a range of different food products to find out how many of them contain additives as well as the different types of additives they contain.

2 Draw up a chart to record your findings, making note of the types of additives used and why they have been used.

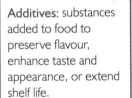

KEY WORD

Additives: substances added to food to preserve flavour, enhance taste and appearance, or extend shelf life.

KEY POINTS: FOOD MODIFICATION AND THE USE OF ADDITIVES

- Manufacturers sometimes add additives during food processing to extend the shelf life of foods or to improve or change their flavour, taste or appearance.
- Preservatives, flavour intensifiers, stabilisers, emulsifiers and colourings are all examples of additives.
- There is concern that some additives may have negative impacts on health, such as causing allergies or being linked to hyperactive behaviour in children.

Exam practice questions

1 Some food products contain additives. Describe the benefits of using additives when manufacturing food products.

[6 markes]

SECTION 6

Cooking and food preparation

This section contains the following chapters:

CHAPTER 10
Factors affecting food choice

Learning objectives

In this chapter you will learn:

- how sensory perception guides the choices that people make, and how taste receptors and olfactory systems work
- the sensory qualities of a range of foods and combinations, and how to set up taste panels for preference testing
- the range of factors that influence food choices, including enjoyment, preferences, seasonality, costs, availability, time of day, activity, and celebrations, occasion and culture
- the choices people make about certain foods according to religion, culture, ethical belief, medical reasons or personal choices
- how to make informed choices about food and drink to achieve a varied and balanced diet, including awareness of portion sizes and costs
- how information about food is available to the consumer, including food labelling and marketing and how this influences food choice.

How sensory perception guides the choices that people make

Sensory perception is the way we recognise flavour in food. Flavour is the sum of all the sensory stimulators, which are shown in Figure 10.1. Our brains receive flavour as a mixture of a food's taste, smell, look and feel in one single sensation. These sensory stimulators have an influence on our food choices.

Taste

Taste is consistently thought of as the major influence on how we select and enjoy our food.

From an early age, taste and familiarity influence our food choices. A liking for sweet foods and a dislike of bitter foods are thought to be an innate part of human nature – i.e. a trait you are born with.

- Taste preferences develop through experience.
- Genetic differences are shown to give different abilities to perceive bitter tastes.
- There are people who are known as 'supertasters'. These people experience taste with far greater intensity than average people. They are more likely to be women and to come from South America, Asia and Africa.

Figure 10.1 Senses involved in sensory perception

How taste receptors work

Taste buds are found on your tongue. Your tongue is covered in tiny nipple or hair-like structures called **lingual papillae** (each one is called a lingual papilla). There are four types of papillae on the human tongue; three of these are associated with taste buds.

Your mouth contains many receptors for taste, and these are found on the upper surface of the tongue, on the soft palate, on the inside of the cheek, in the upper oesophagus (food pipe to your stomach) and on your epiglottis (a flap of cartilage behind the root of your tongue).

The surface of the tongue has tiny holes, or pores, in it. As you chew the food you are eating, the food dissolves in the saliva you produce. It enters the pores and comes into contact with the **taste receptors**. These are located at the top of the taste buds. The taste receptors send information via the nervous system to the brain, and flavours are identified.

The average life of a taste bud is ten days, so they are continually replaced by the body.

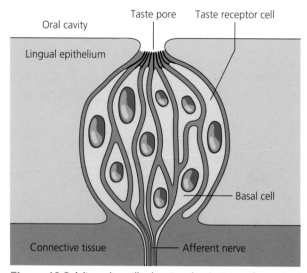

Figure 10.2 A lingual papilla showing the position of taste buds

Figure 10.3 shows the positioning in your mouth and food pipe.

The taste buds detect the five elements of taste:

1   **Sweetness**: all sugary foods.
2   **Sourness**: this is the taste that detects acidity. This taste is found in foods such as lemons, grapes, oranges and sometimes melon.
3   **Saltiness**: any foods containing sodium, sodium chloride or potassium will give a salty taste.
4   **Bitterness**: this is the most sensitive of the tastes, and many people perceive it as unpleasant or sharp. Common bitter foods include coffee, olives, citrus peel, dandelion greens and chicory. Quinine, which is found in tonic water, is also known for its bitter taste.

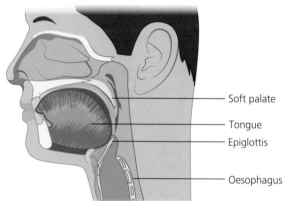

Figure 10.3 The mouth and oesophagus

5   **Umami**: this is a taste that is described as savoury or meaty. It can be tasted in cheese and soy sauce. It is present in fermented or aged foods, and in tomatoes, grains and beans. Monosodium glutamate, a food additive, produces a strong umami taste.

Through the combination of these five elements of taste, we detect **flavours**.

It used to be thought that different areas of the tongue detected different elements of taste, but it is now known that the tongue can detect all of these elements of taste all over the tongue.

Smell

The smell of food contributes a significant proportion of the experience of flavours. Think about the smell from newly-baked bread, or the smell of a roast dinner cooking, for example. People with a reduced or damaged sense of smell cannot experience food tastes with the same level of enjoyment as those who have a full and complete sense of smell. Eighty per cent of what we perceive of the flavour of food is due to smell, and only twenty per cent to just taste.

Smell will increase saliva production, and release gastric juices and insulin into your digestive system.

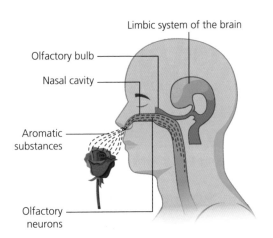

Figure 10.4 **The olfactory system**

How the olfactory system works

The **olfactory system** is found in your nose, and smell detection is by the stimulation of olfactory receptors by particular chemicals in food, which are in gaseous form.

There are tiny hair cells in the olfactory system, which respond to particular chemicals. Humans have about 40 million receptors. (Dogs have about 2 billion receptors!)

The electrical activity from these receptors is passed to the olfactory bulb. This then transmits the information to the limbic section of the brain, where smell is recognised.

If you have a cold, and your nose is blocked, you cannot smell things so well.

The nasal cavity and mouth are connected. That is why smell is connected to taste in flavour perception.

Sight

The saying that we 'eat with our eyes' is true to a certain extent. If the food you are presented with does not look attractive or palatable, you will often not even taste it, and you certainly will not enjoy it as much. Presentation of food is discussed in much more detail in Chapter 11.

Colour is also associated with expected tastes. For example, red foods are usually associated with fruit flavours. If someone gives you a red jelly, you expect it to be strawberry or raspberry flavour, and would be very surprised if it tasted of meat or mint.

Feel

The texture of food is one of the easiest components of flavour to discuss. Creamy, oily, crunchy, chewy or granular are all obvious textures. The part of the brain that deals with aroma or smell (the **olfactory cortex**) is linked to the part of the brain that deals with touch (the **somatosensory cortex**).

Activity

1 To show how much smell is involved in flavour perception, carry out the following experiment:

Take a strawberry or fruit-flavoured sweet. Put it in your mouth and pinch your nose shut. You will taste sweetness and sourness and feel the hard feeling of the sweet. If you release your nose, the smell will travel through your nasal cavity and you will smell strawberry or fruit. You will then taste the strawberry or fruit flavour.

In 2008, a Professor at Oxford University, Charles Spence, won the IG Nobel Prize for proving how important noise is when crunchy snacks are consumed. The study showed that people eating Pringles® thought the crisps tasted stale when they were less crunchy, even though the taste and smell were normal. The study also showed that the louder the crunch, the fresher and crisper the Pringles® were thought to be. This is one reason why crisps are now made to be so noisy, as it makes the brain think they are very fresh!

KEY WORDS

Sensory perception: the way we recognise flavour in food, combining taste, smell, feel and sight.

Taste receptors: these are located at the top of the taste buds and send information via the nervous system to the brain so that flavours are identified.

Lingual papilla: the raised hair-like structures on your tongue that contain the taste buds.

Umami: one of the five elements of taste that has a savoury or meaty flavour.

Olfactory system: the area of your nose that detects smells and aromas.

Olfactory cortex: the part of the brain that deals with smells and aromas.

Somatosensory cortex: the part of the brain that deals with touch.

KEY POINTS: SENSORY PERCEPTION, TASTE RECEPTORS AND THE OLFACTORY SYSTEM

- Sensory perception of food involves smell, taste, sight and feel.
- The tongue is covered in papillae which contain taste buds.
- There are five elements of taste: sweetness, sourness, saltiness, bitter and umami.
- Smell is responsible for 80 per cent of what we perceive as flavour.
- The nasal cavity is connected to the mouth cavity, so the smell of food is combined with the taste.

Exam practice questions

1 Explain why someone who has a reduced sense of smell cannot experience such a heightened sense of taste as someone with a normal sense of smell. [3 marks]

2 Describe how the taste buds in the mouth detect the flavour of the food you are eating. [3 marks]

3 Name the **five** elements of taste and give an example of a food from each one. [5 marks]

Sensory qualities of foods and taste testing

When we talk about the sensory qualities of food we are referring to the look or appearance, the smell or aroma, the taste and the texture (also known as mouthfeel) and if applicable the sound of the food. All of these have been explained earlier in the chapter.

In schools many students will evaluate the **sensory qualities** of food products they have made; this is called sensory analysis and involves the taste testing of the food products. By carrying out taste testing you are able to:

- identify key features of a product
- establish if there are any improvements needed
- determine if a product is suitable/acceptable
- compare similarities or differences in a food product
- test the quality of a food product

Taste tests

There are many different taste tests that are used to evaluate the sensory qualities of food products as well as establish if food products are liked or disliked by other people. Some of these are explained below.

Ratings test

In a **ratings test** taste testers will rate a food product for a particular characteristic such as how crispy it is or how salty it is, as well as give their opinion of the food product by rating it using a scale. This identifies if they really liked it or disliked it. It is called preference rating. An example of a ratings test is shown below. Taste testers are asked to tick in a box to indicate how much they liked the food product.

Sample	Like very much	Like	Neither like or dislike	Dislike	Dislike very much

Taste testers can also be asked to identify an order of preference for food products if they taste more than one. An example would be a food product such as bread made using different flours, where taste testers are asked to rate them from the one they liked the best to the one they felt was the worst. They would write the name of each sample of bread in a box according to how they liked or did not like them. The samples would be given random codes to identify them by.

1 (the best)	2	3	4	5 (the worst)

Profiling test

A **profiling test** would be used to find out what people especially liked about a food product. This type of test would help develop a profile according to a range of sensory qualities for example, chewiness, creaminess, saltiness. Taste testers would be asked to score the food product out of five – one would be the lowest and five would be the highest. For more than one person tasting the product the scores are added up and divided by the total number of tasters to establish the average score. The example shown below is a profile of a lemon sauce tested by five tasters.

Score key: 1 = poor, 2 = average, 3 = good, 4 = very good, 5 = excellent.

Tasting words	Scores awarded by tasting panel					Total	Average
Lemon flavour	3	2	3	4	3	15	3
Colour	2	1	2	1	2	8	1.6
Smoothness	5	5	4	5	4	23	4.6
Consistency	4	4	3	5	4	20	4

Star profile

A **star profile** (it can also be called a star chart) can be used by tasters when they are assessing key sensory attributes of a product such as the appearance, the texture, the taste and aroma. The tasters rate the product according to the attributes and mark their scores on the star profile. The scores are then joined together to create a profile. This can then be used to identify how the product could be improved.

The example shown in Figure 10.5 is a star profile for almond cookies.

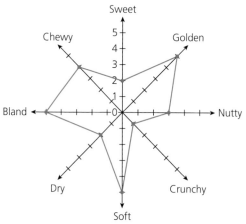

Figure 10.5 A star profile for almond cookies

Setting up tasting panels for preference testing

Food manufacturing industry

The setting up of tasting panels in the food manufacturing industry is very controlled. This is important to achieve reliable results. There are many aspects that are controlled, including:

- the lighting of the area where the testing is to take place
- the temperature of the area
- the use of individual booths to prevent any communication or influencing from other testers
- the use of trained testers
- the coding of food samples with random numbers
- the serving of food samples on identically sized, coloured and shaped plates or containers

- the temperature of the food being served to ensure it is identical for each tester
- the number of samples being presented to the tester
- the cleansing of the palette between the tasting of samples by the testers, by drinking water or eating a plain biscuit/cracker
- the use of clear instructions to give to the tester
- the use of results sheets that are clear and easy for the testers to use.

School classroom

When setting up tasting panels in school it is important that you ask as many people as you can so that you get a range of people's views on the food product. It is also important to ask people who fit into the target group you may be focussing on; for example teenagers, or vegetarians.

Before the tasting session you will need to decide on the type of test that would be best for the tasters and draw up a tasting chart or a profile ready for them to complete as they are tasting the food. You will need to give them clear instructions and try to seat them so that they are not talking or sharing ideas with other testers so that you have fair results. After the tasting session it is important to analyse your results and identify where improvements or changes could be made.

KEY WORDS

Sensory analysis: when the sensory qualities of food products are evaluated and taste tested.

Ratings test: when a food product is rated for a particular characteristic using a scale.

Profiling test: used to find out what people especially liked about a food product to help to develop a profile of a product.

Star profile: used to establish key sensory attributes of a product by marking scores on a profile and joining them up.

KEY POINTS: SENSORY QUALITIES OF FOOD AND TASTE TESTING

- Sensory qualities refers to the look or appearance, smell/aroma, the taste, texture and sound of a food product.
- Sensory analysis involves the taste testing of food products and enables you to establish many important features of a food product.
- Ratings tests, profiling tests and star profiles are examples of taste testing methods that can be used.
- Sensory analysis is carried out by food manufacturers under very controlled conditions.
- Sensory analysis is used in the classroom to collect people's views on a food product and find out where improvements can be made.

Exam practice questions

1 State **two** reasons for carrying out taste testing. [2 marks]

2 Explain the difference between a ratings test and a star profile. [4 marks]

3 List **three** controlled conditions a food manufacturer would have to consider when setting up a tasting panel. [3 marks]

The range of factors that influence food choice

The major determinants of food choice are:

- **Biological:** hunger, appetite and taste.
- **Economic:** cost, income, availability.
- **Physical:** access, skills, education and time available.
- **Social:** family, culture, meal patterns and peers.
- **Attitudes:** knowledge about food and beliefs.

The main influences you need to know about are explored below.

Enjoyment

The enjoyment of food is one of life's greatest pleasures. There are various factors that will influence how much you enjoy your food.

Your taste buds vary in sensitivity at different times. If you are very hungry, your taste buds become increasingly sensitive, so food is tastier. During eating, the taste buds gradually become less sensitive to the food. You may notice that you do not enjoy the food so much after several bites. However, your taste buds can still be excited by another taste. Even if you do not want to finish the main course, you can still be tempted by a dessert!

The enjoyment of food is also influenced by its **palatability**. It is dependent on the sensory properties of the food, such as taste, smell, texture and appearance. Sweet and high-fat foods have a high sensory appeal.

Your mood can influence enjoyment. Happiness will increase enjoyment. The people you are eating with can alter your enjoyment. Whether you feel guilty about eating certain foods will affect how you feel.

There are a few rules that can increase your enjoyment of food:

- Allow yourself to get hungry before you eat a meal; if you are not hungry when you start to eat you will not enjoy it so much.
- The tastiest part of a meal is the first part, when your taste buds are most sensitive, so eat slowly to 'savour the flavour'.
- Keep your meals to a reasonable size, so you can enjoy a dessert after.

Preferences

Humans appear to have an innate ability (i.e. one you are born with) to prefer sweet foods.

When a baby is weaned (starts to eat solid food) the foods given will influence their preferences throughout their lives.

Figure 10.6 Baby eating solid food

Figure 10.7 British asparagus is only available in May and June every year

Figure 10.8a–b A low-income shopper is more likely to choose a value loaf than a bag of apples

It often takes a baby several tries before he or she likes a newly introduced food.

Taste preferences and dislikes for certain foods also develop during experiences. You may feel unwell after eating a certain dish. It may be that the dish caused you to feel unwell, but it also could be totally unconnected with the food. However, you are more likely to associate feeling unwell with the food, and will probably not eat that dish again.

Taste preferences also develop through our attitudes, beliefs and expectations. If your parents do not like certain foods, they will not buy, cook or serve them for you; this means you will not have the experience of trying that food, and are more likely to say you do not like it because your family does not like it.

Seasonality

Fresh fruit and vegetables are grown in the UK, but are only available as freshly grown at certain times of the year when the crop ripens. An example of this is British asparagus, which is only available from the start of May and through June.

If someone wishes to eat freshly grown British fruit and vegetables rather than imported fruit and vegetables from abroad, or produce that has been stored for several months, they will be constrained by the **seasonality** of the fruit and vegetables.

Nowadays, most foods are available all year round, due to the fact that they are imported from other countries. British consumers expect to be able to buy any food they want at any time of the year. This means that many foods have travelled long distances to get to the shelves in the shops and supermarkets. (The impact of this on the environment, and food miles, are discussed in Chapter 8.)

Seasonality can also mean the times of year. We eat different foods in summer than in winter. You are more likely to eat lighter dishes, such as salads, in hot weather. We barbecue when it is warm and sunny, eat ice creams and drink cold and iced drinks. In winter, it is more likely that we eat soups, stews and pies to fill us up and keep us warm. We like hot drinks and roast dinners.

Costs

The cost of food is a primary factor affecting food choice. A person's income or the amount of money they have to spend on food will regulate their food choices.

Low income groups are more likely to eat an unbalanced diet that is based on cheaper foods, which tend to be higher in fat and sugar. Studies show that many low-income groups have low or reduced intakes of fresh fruits and vegetables. An example of this could be as follows: when faced with the choice of a value loaf of bread for 50 pence or a bag of apples for £1, the low-income shopper is more likely to buy two loaves of bread which will provide breakfasts of toast for several days for the family, whereas the apples will not fill children up as much as the carbohydrate-based bread.

The reluctance of someone on a low income to try different foods in case they are rejected and thrown away, which will waste precious money, could be another reason to restrict choice and preference.

Access to more money to spend on food does not necessarily automatically make the diet healthier, but extends the range of food that the person is able to afford. This is because of the preferences and experiences of the person with the money.

Activity

Look at the two shopping lists below.

Shopping list number one

- A 'value' white loaf
- A box of own brand cornflakes
- A large bag of own brand porridge oats
- A large container of milk
- A pack of own brand 'butter-like' spread
- A jar of value jam.

Shopping list number two

- A bag of apples
- A bag of oranges
- A large bunch of bananas
- A box or bag of grapes.

1 Using a supermarket's website, check the prices of the food items in each shopping list.

2 Add up the total cost of each shopping list.

3 Compare the total cost of the two shopping lists.

4 Explain why a low-income parent would probably buy the items on the first shopping list rather than the fruit on the second shopping list.

Availability

The availability of food to a shopper will also restrict what they can buy and eat. We have discussed seasonality above, which may restrict availability to a consumer. Accessibility to shops will also make a difference to what a consumer can buy. This is dependent on transport and geographical location.

- If you live near a large choice of supermarkets, own a car and can drive, you will be able to buy everything you want, easily and quickly.
- If you live on a large housing estate, or outside a town and rely on public transport, it will be more difficult for you to buy and transport what you need. You may have to rely on small, local shops which do not stock such a large variety of products as large supermarkets.
- If you live in a village in the countryside, or on a remote Scottish Island, your choice will be much more restricted.

Figure 10.9 Food available in a village shop

Stretch & challenge

Using the internet, investigate whether you can find any information about the health of individuals who live on high-density housing estates in a large city, and do not have access to large supermarkets because of lack of transport.

Write up the information you find.

Time of day

The time of day that you are shopping, preparing or cooking can influence what you choose.

Studies show that if you shop when you are hungry, you are more likely to spend more money and select foods that appeal to you because they are high in fat and sugar, and would immediately make you feel less hungry.

When preparing and cooking foods, meal times will influence what you decide to eat. Breakfast foods are likely to be cereals or a cooked breakfast, lunches could be sandwiches or a cooked lunch, and an evening meal is more likely to be a cooked dish.

Sitting down at a table and eating meals at regular times also will regulate what you choose to eat. If you are served freshly cooked meals at certain times, the type and amount of food will differ from another situation – for example when you are choosing a McDonald's meal to eat when walking down the street.

Late night snacking, or eating on your way home after a night out is much more likely to involve high-fat, high-sugar dishes or products.

What activity you have planned

Your choice of food is also dependent on what you are planning to do during the day. If you are in a rush in the morning, you will skip breakfast, or grab a breakfast bar. You may buy a bacon roll on the way to work or school.

Many families eat breakfast or brunch out at the weekend, when they have time to enjoy it together as a family.

If you are taking part in a sporting activity, your choice of food will reflect the activity. A prolonged activity, like a marathon race or football, rugby or hockey match will mean you need a slow release of energy, so a starchy-based meal will be a good choice. A shorter sprint race will need an instant supply of energy.

Many families have busy lives and do not eat together every day; this impacts on choice of foods, whether ready meals are purchased, or fresh dishes are made and frozen or portioned in the fridge to be reheated.

Figure 10.10 High-fat item of takeaway food

Activity

1. Make a list of your favourite foods.
2. Put these foods into set meal times: what you have for breakfast, lunch, supper and snacks.
3. Which foods are high-fat or high-sugar foods?
4. Do the snacks tend to be less healthy than the items you eat for set meals? Explain why you think your eating habits are like this.

Celebrations, occasions and culture

A celebration of any type often involves a special meal. Birthdays, weddings and anniversaries are often celebrated with a party, which involves buying and preparing special foods and dishes.

Christmas is a huge family occasion when traditional foods like roast turkey and Christmas pudding are eaten.

Many religions have their own celebrations and traditional foods that are eaten at certain times of the year. (Different religions are discussed in Chapter 4 Plan balanced diets.)

Activity

1 Select one of the following religions: Judaism, Hinduism or Islam.

   a) Find one of the celebrations from that religion and investigate which special dishes are cooked for that celebration.

   b) Make a presentation for your class showing the dishes and their special significance.

Figure 10.11 Traditional Christmas dinner

KEY WORDS

Seasonality: the times of year a particular food product is available because of the time it ripens. The food is also usually at its cheapest price at this time. An example is British asparagus, which is only available in May and June, because that is when it is ready to harvest.

KEY POINTS: FACTORS THAT INFLUENCE FOOD CHOICE

- The main factors that influence choice of food are biological, economic, physical, social, your attitudes and knowledge of food, and your beliefs.
- Enjoyment of food is often dependent on how hungry you are.
- Your food preferences are usually established in your childhood.
- Your income and the availability of food will influence your choice of products to buy and eat.
- The time of day and your planned activities will make a difference to your choice of food.
- Celebrations, occasions and culture often require special meals and ingredients to be purchased.

Exam practice questions

1 Explain why someone living in a small village, who relies on a daily bus service to get to the local town, will have less opportunity to buy a large range of food.   [6 marks]

2 Identify the type of snack that someone walking home after a night out with friends might buy. Discuss why they select a product like this to buy.   [4 marks]

The choices that people make about foods according to religion, culture or ethical belief

Religions and cultures

The different religions that you need to know about are discussed in Chapter 4, and Table 4.2 lists all the foods that Muslims, Jews and Hindus are allowed to eat, and those that they must avoid because of their religious rules.

Briefly:

- Muslims eat Halal food that is killed and prepared in a certain way.
- Jews only eat Kosher foods.
- Hindus are mainly vegetarian, but if they do eat meat, they do not eat beef or pork.

Different cultures will celebrate festivals and events that require special foods.

An example of this is shown below.

Passover

Jewish people celebrate Passover.

Passover is celebrated to remember how the children of Israel left slavery behind them when they were led out of Egypt by Moses over 3,000 years ago.

Passover is celebrated for eight days between March and April, depending on the calendar for that year.

On the evening before Passover, a special service called a Seder takes place over a family meal.

A Seder plate is placed on the table which has a lamb bone, a roasted egg, a green vegetable to dip in salt water, bitter herbs made from horseradish and Charoset (a paste made of chopped apples, walnuts and wine).

On the table there are three Mazot (unleavened bread) on top of each other. Four cups of wine are also placed on the table.

- The lamb bone symbolises the sacrifice brought to the temple in Jerusalem on the afternoon before Passover.
- The egg becomes hard when cooked, so symbolises the Jewish determination not to give up while they were slaves.
- The green vegetables dipped in salt water are a reminder of the tears shed by the slaves.
- The bitter herbs symbolise the suffering of the Jews in Egypt.
- The Charoset symbolises the mortar used by Jews to make bricks while they were slaves in Egypt.
- The four cups of wine remind Jews of the four times God promised freedom to the Jews.

This is a festival and celebration that is specific to Jewish religion.

Cultural differences, however, can alter. When people move to a new country they will often adopt the food habits of the local culture.

Ethical beliefs

These are also discussed in more detail in Chapter 4.

Many people become vegetarian or vegan because of ethical beliefs.

Figure 10.12 One of the symbols used to show that a food product is suitable for vegetarians

Some of the reasons include:
- They do not agree with killing animals for human consumption or they do not like the way animals are treated during the rearing and killing processes.
- They do not agree with land being used to raise animals when it can support many more crops for human consumption.
- They do not want to eat the hormones, chemicals and additives in meat.

The choices that people make about foods according to medical reasons or personal choices

Medical reasons

Medical conditions that will influence food choices are discussed in greater detail in Chapter 4.

The medical conditions that require special or restricted diets are:
- **Pregnancy:** there are certain foods that should be eaten to ensure the correct development of the baby, and certain foods that should be avoided.
- **Coeliac disease:** an allergy to gluten that requires careful selection of foods.
- **Type 2 Diabetes:** a condition that is carefully monitored and controlled by diet and insulin injections.
- **Cardiovascular disease:** a condition requiring a lower-fat diet.
- **Coronary Heart Disease:** a condition requiring a low-cholesterol diet.
- **Stroke:** requiring a similar diet to someone with Coronary Heart Disease.
- **Obesity:** requiring lower-calorie foods to help with weight loss.
- **Allergies and intolerances**, such as nut allergy or lactose intolerance that require avoidance of certain foods, or special products.

Personal preferences

Personal choices can be extremely varied and based on upbringing, religion, peer pressure, income and all the other points we have discussed above.

> **KEY POINTS: HOW RELIGION, CULTURE, ETHICAL BELIEF, MEDICAL REASONS AND PERSONAL PREFERENCES AFFECT FOOD CHOICE**
> - People from certain religions and cultures, such as Jewish people, Muslims and Hindus, will have special diets and restricted foods due to their religious laws, influencing their selection and food preferences.
> - Medical conditions will dictate what foods can and cannot be eaten, changing possible choices of foods for sufferers of these conditions.

How to make informed choices about food and drink to achieve a varied and balanced diet, including an awareness of portion size and costs

Chapter 2 in this book details the necessary nutrients needed in a balanced diet that will ensure that your body functions correctly and that you are as healthy as possible.

The main nutrient groups are:

- Macronutrients: proteins, fats and carbohydrates.
- Micronutrients: vitamins, minerals and trace elements.

A balanced diet will contain a variety of foods that provide the necessary nutrients.

Chapter 3 contains information about calculating the number of calories needed to maintain weight by giving details of how to work out Basal Metabolic Rate (BMR), Physical Activity Levels (PAL) and Estimated Average Requirements (EARs).

Chapter 4 looks at government recommendations for a healthy diet, including the Eatwell Guide. It also has information about the nutritional needs of different age groups, religions and people with dietary-related illnesses.

Figure 10.13 A food label showing nutritional content information

All processed food is labelled with information to allow the consumer to make an informed choice of what to eat.

Food labelling is discussed in more detail in the next section.

Chapter 12 looks at developing recipes to cater for specific needs, and how to reduce the cost of foods.

Portion size can be calculated by looking at recipes, which usually give the number of servings for the finished dish, or by checking the packaging of a ready meal, which will state the number of servings per product. It is possible to calculate the total number of calories per portion by adding up the total calorie values of foods used in recipes and dividing by the number of portions you are serving.

All this information can be used to find and plan a balanced diet for individuals.

How information about food is available to the consumer

There are huge numbers of television programmes, websites, articles, magazines and newspaper columns that are dedicated to providing information about food for the consumer.

New information about what we should and should not be eating seems to be released nearly every week, so it can be very confusing as to what is or is not healthy eating.

Government advice is usually well-documented and backed with scientific information, so using the government guidelines will be a dependable source of information. This includes the Eatwell Guide and the recommended intake of nutrients.

Other reliable information can be found on the NHS website (**www.nhs.uk**), from organisations that are particularly dedicated to certain dietary conditions, such as the British Diabetic Association (**www.diabetes.org.uk**) or British Heart Foundation (**www.bhf.org.uk**), and other dedicated charities who will provide suggested eating plans and recipes for specific conditions.

Food labelling

In December 2014, food labelling regulations changed to make manufacturers include further information on food labels by law.

Food labels must have the following information on them by law:
- **The name of the food** – for example 'wholemeal bread' or 'cottage pie'.
- **A best-before or use-by date**:
  - **Best-before** means the food will be in top condition before that date, but is still usable after that date, but may not be in such good condition.
  - **Use-by** means that the food will be unsafe to eat after that date.
- **Quantity information** – this must be in grams, kilograms, millilitres or litres on the labels of packaged food over 5 g or 5 ml. Solid foods packed in liquid must show the drained weight, i.e. the weight without the liquid. You must be able to see the quantity information when you read the name of the food on the label. You do not have to show the weight or volume of foods sold by numbers, e.g. 2 bread rolls, provided you can clearly see the number of items inside the packaging.
- **A list of ingredients** (if there are more than two) – ingredients must be listed in order of weight, with the heaviest listed first. You have to show the percentage of an ingredient if it is highlighted by the labelling or a picture, e.g. 'extra cheese', or mentioned in the name of the product e.g. 'cheese and onion pasty', or normally connected with the name of a product, e.g. the fruit in a summer pudding.
- **Allergens** – if your product contains any of the following allergens you must say so clearly on the label and list them in the ingredients:
  - celery
  - cereals containing gluten – including wheat, rye, barley and oats
  - crustaceans – including prawns, crab and lobster
  - eggs
  - fish
  - lupin
  - milk
  - molluscs – including squid, mussels, cockles, whelks and snails
  - mustard
  - nuts

- o peanuts
- o sesame seeds
- o soya beans
- o sulphur dioxide or sulphites at levels above 10 mg per kilogram or per litre.

- **The name and address of the manufacturer, packer or seller** – this is in order to have the contact details in case there is a problem with the food.
- **The lot number of the food** – this means it can be traced back to the production date.
- **Any special storage conditions** – e.g. store in the refrigerator.
- **Instructions for cooking if necessary**.
- **Country of origin** – if the following products in your food have been imported from outside the EU: veal, beef, fish, shellfish, honey, olive oil, wine, most fruit and vegetables, poultry.
- **A warning if the product contains GM ingredients** unless they are less than 0.9 per cent of the final product.
- **A warning if the product has been irradiated**.
- **The words 'packaged in a protective atmosphere'** if the food is packaged using a packaging gas – this could be in Modified Atmospheric Packaging, for example.
- **Any necessary warnings** – the list below shows the government legislation for which ingredients and chemicals must be listed, and the actual wording that has to be added to each ingredient.

Ingredient	Wording you must use
Allura red (E129)	'May have an adverse effect on activity and attention in children'
Aspartame	'Contains a source of phenylalanine'
Caffeine over 150 mg/l	'Not suitable for children, pregnant women and persons sensitive to caffeine'
Carmoisine (E122)	'May have an adverse effect on activity and attention in children'
Liquorice	'Contains liquorice' (you may need extra wording for confectionery or alcohol containing liquorice)
Polyols	'Excessive consumption may cause a laxative effect'
Ponceau 4R (E124)	'May have an adverse effect on activity and attention in children'
Quinoline yellow (E104)	'May have an adverse effect on activity and attention in children'
Raw milk	'This milk has not been heat-treated and may therefore contain organisms harmful to health'
Skimmed milk with non-milk fat	There's no fixed wording, but you must show a warning that the product is unfit to be used for babies.
Sulphur dioxide over 10 mg/l	'Contains sulphur dioxide (or sulphites/sulfites)'
Sunset yellow (E110)	'May have an adverse effect on activity and attention in children'
Sweeteners	'With sweetener(s)'
Sweeteners and sugar	'With sugar and sweetener(s)'
Tartrazine (E102)	'May have an adverse effect on activity and attention in children'

Table 10.1 Listed chemicals and ingredients

From December 2016, the law states that mandatory nutritional information must be included on food labels.

What other information is on food labels and why?

Many manufacturers choose to put added information on food labels. This is to inform the consumer, but also to attract customers to the product and persuade them to buy that product.

Nutritional information
- Energy per 100 g and per portion.
- Reference Nutritional Intakes (RNIs).
- % RNI is used for fats, saturated fats, sugars and salt.
- Traffic light symbols: traffic light symbols are used to indicate whether a product is high (red), medium (amber) or low (green) in fats, saturated fats, sugars and salt. These also show the amount of nutrients in a portion of food and drink, and the percentage of your RNI that a portion of the product will provide. This allows a consumer to make an informed choice about the product.

Activity

1 Collect a variety of food labels. See if you can find all the necessary information that is required by law on the labels.

2 What other information is included that **is not** required by law? Can you think of reasons why this extra information has been included?

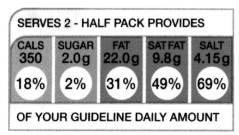

Figure 10.14 Example of traffic light labelling of a product

'May contain'

Some products state that they 'may contain' a product, for example nuts. This is because the manufacturer may not be able to guarantee the product is free from that allergen.

Nutritional and health claims

These are also covered by European regulations, but are a separate regulation to the food labelling laws. To use a health claim on a food or drink, it must be authorised and listed on the European register of claims and it must meet certain conditions. For example, if a food says it is a source of fibre, it must have at least 3 g of fibre per 100 g of product.

Examples of nutrition claims are:
- Sugar-free (must contain less than 0.5 g of sugar per 100 g of product).
- Low fat (must contain less than 3 g of fat per 100 g of product).
- High in fibre (must contain more than 6 g of fibre per 100 g of product).
- Source of vitamin D (must contain at least 15 per cent of the RNI for vitamin D per 100 g of product).

Health claims are claims that suggest there is a relationship between the product and good health. Examples of health claims are:
- Calcium is needed for the maintenance of normal bones.
- Plant sterols have been shown to lower/reduce blood cholesterol. High blood cholesterol is a risk factor in the development of Coronary Heart Disease.

Activity

Why do you think health and nutritional claims have to be regulated by a European Authority?

The health claim of the company must be reviewed by a panel of experts (the European Food Safety Authority) who assess whether the claim is supported by scientific evidence. They then decide whether to authorise the claim and how it should be worded.

E numbers, antioxidants and preservatives

These are listed in the ingredients. They can be checked against lists found on the internet that will identify the names of the chemicals used (**www.ukfoodguide.net** will provide the information you need). The British Nutrition Foundation website (**www. nutrition.org.uk**) also has information on E numbers.

Flavourings, flavour enhancers, sweeteners, emulsifiers and gelling agents

- Flavourings and flavour enhancers are used to add or enhance flavour in a food. Flavourings are not given E numbers, but 'flavouring' will be listed on the ingredients.
- Sweeteners are often used to add a sweet taste in low-calorie products. Common sweeteners include aspartame (E951) and saccharine (E954).
- Emulsifiers allow fat and water to mix, when naturally they would separate, e.g. in mayonnaise. A common emulsifier is Lecithin (E322), which is found naturally in many foods. Stabilisers such as locust bean gum (E410) made from carob beans, help to keep these mixtures from separating again.
- Gelling agents such as pectin (E440), which is a type of soluble fibre present in fruit such as apples, provide texture to a product, and thickeners add body to products such as sauces and soups.

All additives used in the EU must have undergone rigorous safety testing.

Marketing terms

Terms such as 'fresh', 'pure', 'natural' and 'traditional' are often used to describe foods, but do not have legal definitions, and do not really mean anything. The use is checked, so you cannot use 'natural' for a product that contains artificial additives, or 'fresh' for a juice that is made from fruit concentrates.

Vegetarian and vegan labelling

If a food is labelled vegetarian, it should not contain any meat, fish or poultry, or animal-derived additives such as gelatine.

Vegan-labelled products should not contain any animal products.

Made with real fruit or contains real fruit juice

There is no law to say how much fruit should be in a product, so you will need to check the ingredient list to find out how much fruit is contained in the product.

Wholegrains

Refined white flour with a touch of whole wheat added to it can be listed as wholegrain. You will have to check the ingredients list to see how much wholegrain is in the product.

Variations of this include:

- 'Made with wholegrains' – see above description.
- 'Wheat flour' or '100 per cent wheat' – you need to look for *whole* wheat flour.
- 'Multigrain' – this means there are several types of grains, not necessarily wholegrains.
- 'Wholegrain' – check the ingredients and avoid words like bleached or enriched. 100 per cent wholegrain is the best.

Celebrity endorsements

Many companies will pay well-known celebrities to advertise and appear on the packaging of their products. This is used as a marketing tool to persuade people to buy their products.

Cartoons

Lots of products that are aimed at children will use cartoon characters to decorate the packaging of the product. This is to try to attract children to the product, and make them pester their parents to buy the product when shopping.

Free gifts

Lots of manufacturers will offer free gifts such as discounts for entry to theme parks or child-friendly venues on children's products, fitness equipment on products that help you to lose weight, free toys inside the packaging, or any other marketing campaigns that will encourage you to buy a certain product over a rival product.

Claims of 'new' or 'improved' recipes or ingredients

Once a product has been well-established, a manufacturer may change the recipe or ingredients slightly to try to attract new customers.

KEY POINTS: INFORMATION ABOUT FOOD AVAILABLE TO THE CONSUMER

- There are European food labelling laws that require all food labels to have certain information.
- Many manufacturers put additional information on their labels to inform their customers and to attract customers to their products.
- Marketing of foods may include celebrity endorsements or cartoons to promote the product, encouraging adults and children to buy those products.

Exam practice questions

1 List **three** things that are required by law to be on a food label. [3 marks]
2 Explain why it is necessary to put both storage instructions and cooking instructions on a packed food. [4 marks]
3 Discuss reasons why manufacturers put traffic light symbols on their products when it is not a legal requirement. [4 marks]
4 Identify **one** product that you know has been endorsed by a celebrity, or uses a cartoon character to advertise a product, and explain why you think this would encourage someone to buy the product. [4 marks]

Learning objectives

In this chapter you will learn:

■ how to plan and cook a single dish
■ how to plan and cook a number of dishes in one session, producing a plan which shows dovetailing
■ all the skills you need to be able to demonstrate
■ how to prepare different ingredients for a recipe
■ how to cook and present a selection of recipes, including cooking and serving techniques
■ how to work safely and hygienically while working independently
■ how to use sensory descriptors appropriately and correctly.

Planning for cooking a single dish

Many of your practical sessions will involve independent or group work, cooking one recipe. This is going to prepare you and improve your cooking skills, allowing you to 'tick off' all the skills you need (see the list in Table 11.3).

When you are selecting a recipe, there are several factors you need to take into account.

1 Look at the task you have been given

It is very important to read the task carefully to make sure that your selection of recipe, or recipes, fulfils all the criteria.

An example of a task could be: 'To prepare a dessert that is suitable for someone who is lactose intolerant'.

You would either:

● look for a recipe that is designed for someone with lactose-intolerance; or
● you might alter a normal recipe to substitute ingredients that contain lactose for lactose free ingredients.

2 Look at the amount of time you have to prepare and cook the dish

Many lessons in school are only an hour long. This means that by the time you have washed your hands, sorted out your ingredients and found the equipment, you probably only have 50 minutes to actually prepare, cook and clear up afterwards.

You need to select a recipe that can fit this time frame.

The best way to work this out is to write a time plan to see if the recipe fits in.

Table 11.1 shows an example time plan for a Victoria sandwich cake.

Time	Action	Health and safety points
11.00	Wash hands, put on apron. Put oven on to 180°C. Grease and line cake tins.	To kill bacteria and prevent contamination. To pre-heat oven.
11.05	Weigh out ingredients.	Ensure correct amounts are weighed.
11.10	Cream butter and sugar. Add eggs and beat well. Fold in flour.	
11.15	Divide mixture between tins and level tops.	
11.20	Put cake in oven for 20 minutes.	Use oven cloth to prevent burns.
11.25	While cake is cooking, wash up and clean down working areas.	To prevent dirt and bacteria on work surface.
11.40	Check cake is cooked and if so remove from the oven. Leave to cool.	Use oven cloth to prevent burns.
11.50	Spread jam onto one of the cakes and sandwich together.	
11.55	Dust the top of the cake with icing sugar. Wash up last pieces of equipment, and put away.	

Table 11.1 Example time plan for Victoria sandwich cake

By writing a time plan you can see that this recipe just fits in the one-hour lesson that you have.

Writing the time plan also gives you a chance to make sure that you understand the recipe and that you know what you have to do before you start.

Activity

1 Choose a simple recipe from a recipe book or website.
2 Make a time plan for the recipe, using a similar format to Table 11.1. See how long it will take.
3 Now look at the recipe and see if they gave you a time for preparing and cooking. Do your timings match up with the times they gave? If not, why not?

3 Check the cost of the ingredients

You should know whether your family can afford expensive things or not. It is really important that you choose recipes that contain ingredients that your family can afford.

Your school probably has salt, pepper and some herbs and spices, so check before you go and buy a whole jar of something that you may not use again. An example of this is if you are making a curry, which requires several different spices.

There are lot of supermarket websites that allow you to check prices if you are not sure how much things cost.

Activity

Look at the following list of ingredients for a lemon meringue pie.

Lemon meringue pie

250 g plain flour	275 g caster sugar	30 g cornflour
125 g butter	2 large lemons	4 eggs

Use a supermarket website to compare the cost of the ingredients, by selecting 'basic' own brand ingredients for one price, and more expensive branded ingredients for another price. Compare the costs of the two.

(Remember, you will have to work out the exact quantities you need. Some of the items you buy will be in large amounts, for example the flour will be in a 1.5 kg bag, but you only need 250 g, so you will have to work out the cost of 250 g of flour.)

4 Cooking the dish

- Make sure you follow the time plan that you have written. If your time plan is in the correct order, it will be much easier to cook the dish correctly. Have you turned the oven on to pre-heat?
- Make sure you taste the food as you go along, to check that the seasoning is correct.
- Does the dish need to be glazed so it turns a nice golden brown colour in the oven? An example of when this might need to be done is when making a pastry pie or sausage rolls.
- If you are making a cake, is the mixture level in the tins? Is the oven on the correct temperature?

5 Think about the presentation of the dish

Some of the marks for your practical work are for how attractive and well presented the dish is.

We all 'eat with our eyes' before we taste the dish. An attractive dish is much more likely to be enjoyed by people.

- Think about any garnishes you can add to make the dish more colourful.
- Does the dish need a glaze?
- Have you chosen the correct-sized dish for the amount of ingredients?
- Have you dished up a portion onto the plate neatly?

6 Evaluate your dish using sensory descriptors

Once you have finished your dish, get into the habit of looking at it critically and ask other people to do the same.

Ask yourself questions: Does it look attractive? Why or why not?

Use colour, appearance and smell to start with, for example, 'my cake looks golden brown; it is well risen. It smells sweet but with a hint of egg smell'.

Then taste it: for example, 'my dish is soft and sweet with a sugary crust'.

Figure 11.1 A beautifully presented and garnished dish

Think of descriptive words, so that someone who is not in the room with you can imagine how it looks, smells and tastes.

At the end of this chapter, Table 11.4 gives a list of descriptive words you can use.

7 Finally, is there any way you could improve this dish if you made it again?

Your dish may have turned out exactly as you wanted. It may look amazing.

However, you may think it could look much better, so ask yourself: 'What would I do differently next time? Could I improve the appearance? Could I add a garnish for colour?'

Finally, make some suggestions to improve the dish next time you cook it.

Write everything up, and present your completed report to the rest of your class.

Ask the other students to mark your work, giving feedback to you.

Practical activity

1   Select a recipe to cook that fulfils a task you have been given. Check that the recipe meets the brief.
2   Make a time plan for your dish.
3   Use a supermarket website to check the cost of ingredients. Check with your teacher to see if any extra herbs or spices are in the school store cupboard.
4   Cook the dish, following your time plan.
5   Evaluate your dish, and ask your friends to do the same.
6   Finally, make some suggestions to improve the dish next time you cook it.
7   Write everything up, and present your completed report to the rest of your class.
8   Ask the other students to mark your work, giving feedback to you.

Suggestion: Keep a folder with successful recipes and time plans so you can use these for your assessments later in your course.

KEY POINTS: PLANNING FOR COOKING A SINGLE DISH

- Check that your selected recipe fits the task and the amount of time you have to cook.
- Always think about the cost of the recipe.
- Plan the presentation and garnishes to make the dish look attractive.
- Evaluate your dish using sensory descriptors, and think of ways you could improve it next time.

Exam practice questions

1   Give **two** reasons why it is important to write out a time plan for your cooking sessions.   [2 marks]

2   Name **three** things you have to think about when selecting a recipe for a given brief.   [3 marks]

Planning for cooking a number of dishes in one session

Once you have become confident at planning and cooking single dishes, you will move on to cooking more than one dish at a time.

Your non-examination assessment will require you to produce several dishes at the same time.

If you can start to practise doing this, you will be much more confident as time goes on. You will also increase the speed at which you work.

Some of the dishes could be made using the same basic recipe. For example, you could be asked to produce a series of dishes using shortcrust pastry. You would make a large batch of pastry, divide it into several portions and make something different from each portion, showing the versatility of the pastry.

Other tasks may ask you to prepare a selection of dishes suitable for one occasion, like a celebration or three-course meal. You would plan and cook those dishes, and serve them all up at the same time at the end of the cooking session.

When planning to cook more than one dish at a time, you need to think about the following points:

1 Look at the task you have been given

Just as when you were choosing one dish, you need to make sure that the selection of dishes you choose fit the task.

2 Look at the amount of time you have to prepare and cook the dishes

You will not usually be expected to cook more than one dish in a short lesson. The amount of time given to cook a selection of dishes will be longer. However, you need to make sure that all the dishes you have chosen fit in the time you have.

You will need to make a detailed time plan, to check it is possible to make them all.

As you are more confident, you should be a bit faster at techniques, and you should be able to estimate the amount of time you take for each dish.

Your time plans will need to show **dovetailing**. This is when your plan shows that you are dealing with more than one thing at a time – for example, 'while the pasta is cooking, I will make the sauce'. This shows the examiner that you are able to plan carefully and cope with cooking more than one dish at a time.

What must I consider?

1 Choose the dish that takes the longest to cook, and make that first.
2 Anything that has to set in the refrigerator should be made second.

3 Make sure you are checking anything in the oven, so put a reminder for yourself in your time plan when the cooking time is nearly complete.

4 Allow time for clearing up in between dishes.

5 Serve everything up at the same time at the end of the session.

The time plan below shows dovetailing. You can use words and phrases such as 'while the pastry is baking blind, make the filling'.

The task: make a selection of dishes suitable for a buffet lunch.

The time plan is for two hours, making the following dishes:

- Mushroom, pepper and onion quiche.
- Cheese scones, with cream cheese and chilli jam.
- Rolled smoked salmon pinwheel sandwiches.
- Pizza slices.
- Mixed green salad.

Time	Action	Health and safety points
11.00	Wash hands, put on apron, get equipment out.	To kill bacteria and protect clothing and food.
11.05	Make shortcrust pastry for the quiche: • Weigh ingredients, rub in fat and flour. • Add water, mix to a dough. Wrap pastry in cling film and put in fridge to rest for 10 minutes. Turn oven on to 180°C.	Check that equipment is clean. Weigh ingredients carefully. To preheat oven.
11.10	While the pastry is resting, make dough for pizza: • Weigh out flour; add yeast and salt. • Measure and add water; mix to a dough. • Knead for 5 minutes. Cover and put in a warm place to prove.	
11.20	Grease quiche dish. Take pastry from fridge, roll out and place in quiche dish. Prick all over with a fork; place greaseproof paper and baking beans. Put in oven for 15 minutes to bake blind.	Use oven cloth to prevent burning.
11.30	Wash all equipment and wipe down surfaces.	To prevent dirt and contamination.
11.35	Peel and slice onion, de-seed and slice pepper, slice mushrooms. Mix eggs, milk and seasoning in a jug. Grate cheese.	
11.45	Remove quiche case from oven, take out baking beans and return to oven for 5 minutes.	Use an oven cloth to prevent burning.
11.50	Remove quiche case again; place sliced vegetables in the case; pour egg mix over and finish with the cheese. Place back in the oven for 30 minutes.	Use an oven cloth to prevent burning.
11.55	Knock back and roll out pizza dough onto a greased baking sheet. Spread with tomato puree. Slice tomatoes; place on top. De-seed and slice the yellow pepper; place on top with the olives. Cover with grated cheese. Place pizza a in warm place to prove again for 15–20 minutes.	Use an oven cloth.

12.05	Wipe down working area.	
12.10	While the pizza and quiche are cooking, make scone dough: • Grate cheese. • Sieve flour, rub in fat, add grated cheese and milk to make a dough. • Roll out, cut out scones and place on baking sheet. • Brush with beaten egg.	Check equipment is clean.
12.20	Remove quiche from oven. Increase oven temperature to 200°C Bake pizza–20 minutes. Bake scones–10 to 15 minutes	Use an oven cloth.
12.30	Make sandwiches: Spread horseradish cream on slices of bread. Top with smoked salmon. Cut off crusts and roll up. Slice diagonally and put on a plate. Garnish with pieces of dill and tomato.	Take care with sharp knife.
12.35	Remove scones from oven and place to cool.	Use an oven cloth.
12.40	Remove pizza from oven and place to cool.	Use and oven cloth.
12.45	Split scones, spread with cream cheese and top with chilli jam. Slice pizza and place on serving dish.	Take care with sharp knife.
12.55	Serve all food. Wash and clear up.	

Table 11.2 Time plan showing dovetailing

You can see that you are aware of all the dishes that are cooking, resting and proving and are dealing with them in the correct order.

Activity

Create a dovetailed time plan for a selection of dishes that would fit the buffet lunch task above. Remember to consider all the points listed above the task.

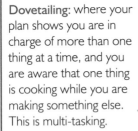

KEY WORD

Dovetailing: where your plan shows you are in charge of more than one thing at a time, and you are aware that one thing is cooking while you are making something else. This is multi-tasking.

KEY POINTS: PLANNING FOR COOKING A NUMBER OF DISHES IN ONE SESSION

- Check your selection of dishes fits the task and can be prepared in the time you have for the cooking session.
- Make sure you are making the dish that takes the longest time to cook first.
- Make sure any dishes that need to chill or set have enough time in the fridge or freezer.
- Show dovetailing in your time plan, so that the examiner can see that you are aware of dishes in the oven, or that you are cooking two or three things at the same time.

Skills you need to be able to demonstrate when preparing ingredients and cooking and presenting dishes

There are 20 different skill groups in your GCSE. You may already have learned to do some of them in your lessons, or at home.

The following table lists all the skills you need to show during your GCSE course.

Skill number	Skill group	Techniques that you need to demonstrate
1	Knife skills	Be able to: • fillet a chicken breast • portion a chicken • remove fat and rind from meat • fillet fish • slice raw and cooked meat or fish and alternatives such as tofu and halloumi, evenly and accurately • cut and slice fruit and vegetables using the bridge hold and claw grip • peel, slice, dice and cut fruit and vegetables into evenly-sized pieces.
2	Prepare fruit and vegetables	Be able to: • mash, shred, scissor snip, scoop, crush, grate, peel and segment fruit and vegetables • de-skin, de-seed, blanch, shape, pipe, blend, and juice fruit and vegetables • prepare garnishes. All of these must be done while demonstrating the skills of controlling enzymic browning (see Chapter 6 for a description of this) and spoilage, and preventing food poisoning.
3	Prepare, combine and shape	Be able to: • roll, wrap, skewer, mix, coat, and layer meat, fish and alternatives • shape and bind wet mixtures, such as fish cakes and meatballs. All of these must be done while demonstrating the prevention of cross-contamination, and showing that you can handle high-risk foods safely.
4	Tenderise and marinate	Be able to: • show how acids denature proteins by using a marinade • show how marinades add flavour and moisture while preparing meat, fish, vegetables and alternatives.
5	Select and adjust a cooking process	Be able to: • select and adjust the cooking process and length of time to suit the ingredient. An example of this is that your meat or fish may be a thinner cut, so will need less time than the recipe says.
6	Weigh and measure	Be able to: • weigh and measure out solid and liquid ingredients accurately.
7	Preparation of ingredients and equipment	Be able to show you can: • grease, oil, line and flour evenly. You need to show you can do this to make sure your finished product looks as good as possible.
8	Use of equipment	You need to show you can use the following equipment safely and accurately: • blender • food processor • mixer • microwave oven.

9	Water-based cooking methods using the hob	You need to show you know how to:
		• steam foods (e.g. vegetables)
		• boil and simmer foods (e.g. vegetables)
		• blanch foods (e.g. vegetables)
		• poach foods (e.g. salmon or eggs).
10	Dry heat and fat-based cooking methods using the hob	You need to show you know how to:
		• dry fry (e.g. beef mince)
		• pan or shallow-fry (e.g. a piece of fish)
		• stir-fry (e.g. a mushroom stir-fry in a wok).
11	Using the grill	You need to be able to show that you can use the grill to:
		• char (this means to blacken the surface of a food such as a pepper)
		• toast and grill.
		You can do this with a range of foods such as vegetables, meat, fish or alternatives such as halloumi cheese, or seeds and nuts.
12	Using the oven	You need to show you know how to:
		• bake (e.g. making cakes or biscuits)
		• roast (e.g. a roast piece of meat)
		• make a casserole or use a tagine (e.g. a chicken casserole)
		• braise (e.g. braised beef).
13	Make sauces	You will need to show you know how to show that starch gelatinisation has taken place by doing the following things:
		• make a roux sauce
		• make an all-in-one sauce
		• make an infused sauce such as a velouté or béchamel sauce.
		This will demonstrate understanding of how liquid/starch ratios affect how thick the sauce is, and how conduction and convection currents work to cook the sauce. You also need to show that you understand the need to stir the sauce continually to prevent sticking, burning and lumpiness.
		• Make a reduction sauce, where the sauce simmers to thicken it. Examples are curry sauce, pasta sauce and meat sauce (to include meat alternatives, such as Quorn™).
		This will show that you understand that allowing the sauce to simmer makes the liquid evaporate, and thickens the sauce.
		• Make an emulsion sauce such as mayonnaise or hollandaise.
		This will show you know how to make a stable emulsion.
14	Set a mixture by removing heat (gelation)	Be able to show you know how to:
		• use a starch to set mixture on chilling for desserts such as cheesecake or custard.
15	Set a mixture by heating (coagulation)	Be able to show you can:
		• use protein to set a mixture when it is heated (e.g. using the denaturing of the protein in eggs to set the inside of a quiche, or making choux buns).
16	Using raising agents	Be able to demonstrate the following techniques:
		• use egg (colloidal foam) as a raising agent, by creating a gas-in-foam product (e.g. whisked sponge or meringues)
		• use chemical raising agents such as self-raising flour, baking powder and bicarbonate of soda
		• use steam in a mixture as a raising agent (e.g. choux buns).
17	Make a dough	Be able to show you know how to use the technical skills of:
		• shortening (e.g. using fat to coat flour to shorten gluten strands in shortcrust pastry)
		• gluten formation (e.g. kneading bread)
		• fermentation (e.g. when you leave bread to prove).

18	Shaping and finishing a dough	Be able to show you know how to:
		• roll out pastry accurately
		• use a pasta machine
		• accurately use pastry to line a flan ring or dish
		• create layers (e.g. using puff pastry to make palmiers or vanilla slices)
		• allow a dough to prove or to rest in the fridge to relax it
		• glaze and finish dough by:
		  ○ piping choux pastry (e.g. éclairs)
		  ○ egg-wash bread rolls to glaze
		  ○ make homemade pasta
		  ○ make flat breads, pinwheels, bread rolls, pizza or calzone.

19	Test for readiness	Be able to show you know how to use all of these methods to see if a dish or food is properly cooked: • use a temperature probe (e.g. to see if the inside of a piece of meat has reached the correct temperature to kill bacteria) • use a knife or skewer (e.g. to see if a fruit cake is cooked in the centre) • check if the 'bite' is correct (e.g. if pasta is cooked) • check if the colour is correct (e.g. a pie crust is sufficiently browned) • if it sounds cooked (e.g. tapping the base of bread to see if it sounds hollow).
20	Judge and manipulate sensory properties	Be able to show you know how to: • taste and season during the cooking process to ensure the food tastes good • alter the taste and aroma by using herbs, spices, pastes, jus and reductions • change texture and flavour by using browning (**dextrinisation**) and glazing. Also by adding crusts, crisp and crumbs to make a product crunchy • present and style food by using garnishes and decorative techniques (such as piping), to improve the aesthetic qualities (how the food looks) • portion and present your dishes properly.

Table 11.3 List of required skills

Activity

1. Ask your teacher to photocopy the list in Table 11.3 for you.

   Look down through the list:

   a) Tick off all the skills you already know, or have covered in your lessons. (There will be many you already know how to do.)

   b) Highlight the ones you have not covered already.

   c) Stick the list into your exercise book or folder, and as you cover each skill mark it off.

   Plan your practical sessions so you can cover the skills you need to.

2. Find two recipes that show the following combination of skills.

   a) Recipe number one: knife skills, weighing and measuring, preparing fruit and vegetables, tenderise and marinade, dry heat and fat-based cooking methods using the hob.

   b) Recipe number two: use of equipment, using a mixer, weighing and measuring, making and shaping a dough, using the oven, testing for readiness.

   Write out your recipe in stages and say where each of the required skills is used.

KEY WORD

Colloidal foam: when two ingredients are mixed together to form a foam, e.g. eggs and sugar.

Preparing ingredients

Skill 1: Knife skills

You will use different-sized knives for different skills. The main knives you will use are:

- vegetable knife
- cook's or chopping knife
- palette knife
- filleting knife
- vegetable peeler.

Figure 11.2 Different types of knives

The **vegetable knife** will be used for small fruit and vegetable preparation, including topping and tailing vegetables, peeling certain fruit and vegetables that have thick skins, and cutting these fruits and vegetables using the **bridge hold**. You should have been taught this when you began your food lessons.

To carry out the bridge hold:

1 Place the fruit or vegetable on a chopping board.
2 Make a bridge over the fruit or vegetable with your hand; your fingers should be on one side and your thumb on the other.
3 Guide the knife under the bridge and cut the fruit or vegetable in half.
4 Use the same technique to cut into quarters if needed.

The **cook's knife** is used to carry out the **claw grip**, which is used for larger fruit and vegetables, such as cabbage.

To carry out the claw grip:

1 Put the fruit or vegetable onto the chopping board.
2 Make a claw with your hand by partly curling your fingers together, resting on top of the fruit or vegetable.

Figure 11.3 The bridge hold

> ### KEY WORDS
>
> **Bridge hold:** the safe way to hold small fruit and vegetables while cutting them into pieces, by making your hand into a bridge across the vegetable or fruit and cutting between your fingers and thumb.
>
> **Claw grip:** the safe way to hold fruit and vegetables when slicing them into small pieces, by making a claw of your fingers to guide the knife.

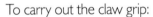

3 Tilt the knife and slice through the fruit or vegetable, using your fingers as a guide.
4 Slide your fingers back, keeping your grip on the fruit or vegetable and continue slicing carefully.

The **filleting knife** will be used for meat and fish. You will need to know how to:

- fillet round and flat fish
- portion a whole chicken correctly, removing the breast fillets and legs
- remove fat and rind from meat
- slice raw and cooked meats evenly and accurately.

Figure 11.4 The claw grip

Step-by-step guide: filleting a round fish

Step 1: Put the scaled fish on a chopping board and, using scissors, trim off the fins by the head on each side, and any fins that run along the top and on the underside of the fish.

Step 2: Remove the head and clean thoroughly.

Step 3: Use a filleting knife to remove the first fillet by cutting along the backbone from head to tail. Keep the knife close to the bone.

Step 4: Reverse the fish and remove the second fillet, this time cutting from tail to head.

Step 5: Trim the fillets to neaten them and skin them using the filleting knife.

Step-by-step guide: filleting a flat fish

Step 1: Place the fish flat on the board and make an incision down the backbone from head to tail, using the filleting knife, along the line that is exactly half way and marks the backbone. Cut round the gills and backbone.

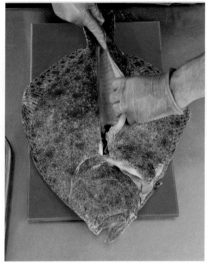

Step 2: Keeping the knife almost parallel to the work surface, carefully slide the knife against the bones underneath, slicing through the flesh and lifting the first fillet away from the fish bones carefully.

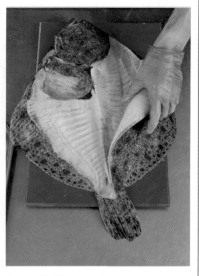

Step 3: Repeat the process to remove the second fillet.

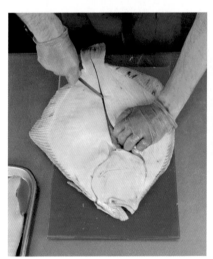

Step 4: Turn the fish over and repeat steps 1 to 4 as above to remove the third and fourth fillets.

Step 5: Hold the fillet firmly at the tail end and cut the flesh as near to the skin as possible, keeping the knife parallel to the work surface, gripping the skin and sliding the knife back and forwards to remove the flesh from the skin.

Step 6: Trim the fillets neatly.

Step-by-step guide: portioning a whole chicken

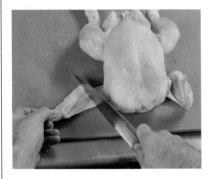

Step 1: Start by removing the legs and thighs. Cut through the skin between the thigh and the body. Bend the leg back so it is lying flat.

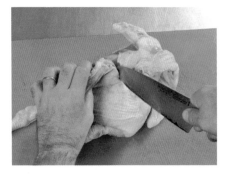

Step 2: Twist the thigh and leg back so the bone comes out of the joint and cut through it to separate the leg from the breast. Do the same on the other side. Cut through the skin to detach both legs completely.

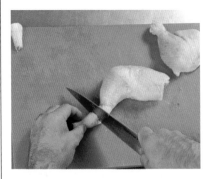

Step 3: Cut through the joint at the base of each leg to remove the feet.

Step 4: Now separate the leg and thigh. Place the leg skin side down ad cut through the joint with the knife. Do the same with the other leg and thigh. You now have four portions.

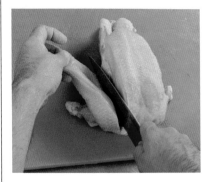

Step 5: Now remove the breast and wings. Cut along the side of the breastbone and then down through the ribcage keeping close to the bone. Then keeping the wing and breast in one piece, cut through the joint at the bottom. Do exactly the same on the other side.

Step 6: Cut through the centre of each breast, cutting the breast, and leaving a third of the meat attached to the wing. Cut the other breast and wing in the same way. You now have eight portions.

Step-by-step guide: filleting a chicken breast

Step 1: Place the chicken breast skin side down on the chopping board and remove all the fat from the breast.

Step 2: Turn the breast over and remove the tenderloin and membrane that connects the tenderloin, as shown in the picture.

Step 3: Pull the skin off the chicken breast.

Step 4: Lay the chicken flat on the chopping board. Place your non-cutting hand flat on the chicken breast. Very carefully begin cutting halfway through the chicken horizontally, keeping your knife parallel with the chopping board. Cut the breast into even-sized fillets.

Figure 11.9a–b How to chop herbs

The **chopping knife** is used to chop, slice, shred and dice vegetables, or chop herbs.

You chop herbs as follows:
1. Gather the herbs together in a ball in your fingertips.
2. Using the claw grip, cut through the herbs a few times.
3. Now place the chopping knife on top of the partly sliced herbs, hold the tip of the blade against the chopping board with the flat of your non-knife hand and finish chopping the herbs by rocking the blade up and down over them until they are finely chopped as shown in the second picture.

This knife will also be used to slice, dice and cut.

WJEC EDUQAS GCSE Food Preparation and Nutrition

Step-by-step guide: slicing strips (julienne)

 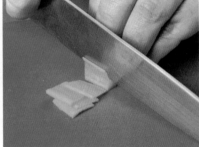 

Step 1 Cut the vegetables into 2 cm lengths.

Step 2 Cut the lengths into thin slices.

Step 3 Cut the slices into thin strips.

Step by step guide: dicing

Step 1 Cut the vegetables into convenient-sized lengths.

Step 2 Cut the lengths into 2 mm slices.

Step 3 Cut the slices into 2 mm strips.

Step 4 Cut the strips into 2 mm squares

Step-by-step guide: cutting batons

Step 1 Cut the vegetables into 1.5 cm lengths.

Step 2 Cut the lengths into 3 mm slices.

Step 3 Cut the slices into batons (3 mm x 3 mm x 18 mm).

Practical activity

1 Show that you can fillet a flat fish, such as a lemon sole, dab or plaice.
2 Use the fillets to make a recipe of your choice. (If you coat the fillets in flour, egg and breadcrumbs before you shallow-fry them, you will be covering many skills in this cooking session.)

Skill 2: Prepare fruit and vegetables

Most of the skills you need to show are ones you will have carried out already.

You will have mashed potatoes, crushed garlic, shredded cabbage or lettuce, grated and peeled fruit and vegetables.

You may have de-skinned tomatoes using the following method:

1 Boil a kettle of water.
2 Place the tomatoes in a heat proof bowl. Cover with boiling water. Leave for two minutes.
3 Carefully remove the tomatoes and place in cold water. You should find that the skin peels off very easily.

Figure 11.13 Peeling tomatoes using boiling water

You should have de-seeded vegetables and fruit. For example, taking seeds out of a whole pepper, or removing seeds from an apple.

You have probably used a blender and a lemon squeezer to juice an orange or lemon. You may have used a juicer.

Scissor snipping is shown in Figure 11.15.

Blanching is when the fruit or vegetable is put into boiling water for two minutes, and it is then plunged into iced water to stop the cooking process. This prevents the enzymes that destroy the vitamins and discolour the fruit or vegetables from being activated (this is called **enzymic browning** – you will have learned about this process in Chapter 6). Vegetables are blanched before they are frozen, so that they do not discolour in the freezer.

Figure 11.14 Removing seeds from a whole pepper

The process is also carried out to allow vegetables to be partially cooked before serving. They can then be reheated without losing colour, so they look more attractive.

The other way to slow down enzymic browning is to put lemon juice on the cut surface to stop the enzymes discolouring the fruit or vegetables.

(Working safely when preparing fruit and vegetables, and preventing food poisoning, is discussed later in this chapter.)

KEY WORDS

Blanch: vegetables or fruit are put into boiling water for two minutes and then plunged into iced water. This helps prevent enzymic browning and partly cooks the vegetables or fruit.

Figure 11.15 Scissor snipping

Figure 11.16 Blanching vegetables

Figure 11.17 Vegetable skewers

Skill 3: Prepare, combine and shape

Many ingredients can be manipulated or mixed with other ingredients to create different-shaped results.

You may have made fishcakes, or prepared meat, fish or vegetable skewers. These would cover many of the skills needed in this section, including skewer, mix, shape and bind wet mixtures.

Practical activity

Make fishcakes using the following recipe.

Easy-to-make fishcakes

Ingredients

400 g fish (e.g. salmon, cod or mackerel)

5 large potatoes

20 g butter

1 onion

1 egg

25 g flour

Salt and pepper

Figure 11.18 Fishcakes

Method

1　Peel and cut up the potatoes into chunks and boil for 20 minutes until tender.
2　While the potatoes are boiling, poach the fish in a little water until cooked and flaky.
3　Chop the onion finely.
4　When the potatoes are soft, drain and mash with the butter.
5　Drain and flake the fish, and add to the potatoes with the chopped onion and egg. Mix well. Season with salt and pepper.
6　Form the mixture into fishcakes, dip the fishcakes into the flour.
7　Fry in a little oil for eight to ten minutes until golden brown on the underside, then turn and fry the other side for five minutes until golden brown.

You could have coated the fish with egg and breadcrumbs or batter before you cooked it.

Practical activity

1　Find a recipe for fajitas.
2　Make the recipe.
　　Making fajitas will show that you have rolled and wrapped ingredients.
3　Tick off any other skills from the boxes in Table 11.3 that you have covered while you were making the fajitas.

(Preventing cross-contamination and handling high-risk food safely when preparing, combining and shaping ingredients are covered later in this chapter.)

Figure 11.19 Coating fish with egg and breadcrumbs

WJEC EDUQAS GCSE Food Preparation and Nutrition

Skill 4: Tenderise and marinade

In Chapter 6 we looked at the process of **tenderising**, by hitting meat with a meat mallet, and **marinating**, by soaking meat, fish and alternatives in a weak acid solution. These processes will **denature** the proteins, making the proteins easier to digest.

If you are barbequing meat, you will need it to be tender, so marinating it before you barbecue will do this.

Skill 5: Select and adjust a cooking process

The instructions given in a recipe for the length of cooking time will be an estimate, based on a particular size or portion of meat, fish or alternative protein source, such as Quorn™ or halloumi cheese.

The size of the portion will almost certainly vary. Therefore, you need to be able to adjust the cooking time to make sure the portion is not undercooked or burnt.

You will need to check during the cooking time to see if a meat portion is cooked by either using a meat thermometer, or by inserting a knife to see if any blood still leaks out.

Fish takes much less time to cook, so this will need to be checked carefully by gently poking a fork into the flesh. It should be opaque and flake away from the fork.

Skill 6: Weigh and measure

Using weighing scales and measuring equipment for liquids accurately will help you to ensure that your recipes are successful.

You will need to be able to use teaspoon, dessertspoon and tablespoon measurements, knowing the difference between a heaped spoonful and a level spoonful of ingredients. Your school may have measuring spoon sets.

Figure 11.20 Perfectly cooked white fish

Figure 11.21 A heaped table spoonful and a level table spoonful of an ingredient

Skill 7: Preparation of ingredients and equipment

In order to prevent food sticking to tins it may be necessary to grease, oil, line or flour cake tins or baking sheets before you use them.

Figure 11.22 Greased and lined cake tin

Activity

Look back through your work and identify all the times you have used the equipment listed in skill 8. Tick these off on your photocopied sheet of skills, making a note of the names of the recipes when you have used these pieces of equipment.

This is the evidence that you have used and understood these skills.

Figure 11.23 A steamer

Skill 8: Use of equipment

Using blenders, food processors, food mixers and microwave ovens are part of the preparation or cooking process for many recipes. Using these items of equipment should not be difficult to incorporate into your practical work.

Cooking a selection of recipes

The following skill requirements involve cooking techniques.

Skill 9: Water-based methods using the hob

The four water-based methods of cooking are steaming, boiling and simmering, blanching and poaching.

Steaming

This is the cooking of prepared foods by steam (moist heat) under varying degrees of pressure.

There are two methods of steaming: low pressure and high pressure.

High-pressure steaming is carried out in a pressure cooker, where the cooker is sealed, with a weight on top, to increase the pressure inside the steamer. The food cooks faster under pressure.

In low-pressure steaming, food may be cooked by indirect or direct contact with the steam.

- **Indirect contact** is when the food is placed in a perforated container above the boiling water, and the steam rises through the perforations and cooks the food. This is when you steam vegetables or fish.
- **Direct contact** is when you place a container into the boiling water. An example of this is cooking a steak and kidney pudding.

Boiling and simmering

This is when you cook foods directly in boiling or simmering water in a saucepan. Many vegetables and fruits are cooked in this way. You also use this method of cooking for potatoes, pasta, rice and stews.

Blanching

This was shown in skill number 2 above. Vegetables are blanched before freezing and to partly cook them so that they can be reheated and served later.

Poaching

This means to cook something in liquid with a temperature ranging from 60°C to 82°C (140–180°F). Poaching is typically reserved for cooking very delicate items like eggs and fish. But other foods like chicken can be poached, along with some vegetables.

Skill 10: Dry heat and fat-based methods using the hob

The three methods in this skill group are dry frying, pan or shallow-frying and stir-frying.

Dry frying

This is when you place an ingredient, such as mince or bacon, which contains fat, into a dry frying pan. You heat the pan gently until some of the fat in the meat begins to melt and this then cooks then food.

An example of when this can be used is when cooking beef or lamb mince for the base of a cottage or shepherd's pie.

Pan or shallow-frying

This is when a small amount of fat or oil is put into a frying pan, the oil is heated and then the food is placed in the hot fat and cooked.

Stir-frying

This is usually done in a wok. Many Asian countries use this method of cooking. It uses very little oil. The wok is heated to a high temperature and the food is added and cooked quickly, as it is shredded into thin pieces.

Figure 11.24 Fish being shallow-fried

Activity

1 Check through your recipe list and identify when you have used water-based or dry heat and fat-based cooking methods.
2 Note these against your list of skills.

This is your evidence that you have used and understood these methods of cooking.

Skill 11: Using the grill

You can use the grill for meats, fish, vegetables, and alternative sources of protein like halloumi cheese. You can also toast seeds and nuts under the grill.

You need to show that you can char vegetables. An example of this is when you blister the skin of a pepper before you take the skin off.

Skill 12: Using the oven

There are four different skills involved that you need to demonstrate: baking, roasting, casseroles or tagines, and braising.

Baking

Baking cakes and biscuits will show this skill.

Roasting

Meat, fish or vegetables can be roasted.

Casseroles or tagines

Casseroles are cooked at a low temperature for a longer time, as they are often made with cheaper cuts of meat that require long, slow cooking to tenderise them.

A tagine is a dish from North Africa, which is named after the pot in which it is cooked.

Figure 11.25 Peppers charring under a grill

Figure 11.26 A tagine

Tagines are usually cooked on an open fire, but can be cooked in an oven. The lid traps the condensation as it rises from food as it cooks, and the moisture filters back into the dish.

Braising

This is when the food, usually a cheaper cut of meat (e.g. brisket), is lightly fried before it is put into a casserole dish and stewed in a liquid. The liquid often comes up to halfway up the joint of meat, so the top turns brown and crisp.

Skill 13: Making sauces

There are three main types of sauce you need to know about, and demonstrate that you can make them. These are blended white sauces, **reduction sauces** and **emulsion** sauces.

Blended white sauces

Blended white sauces show **gelatinisation**. (Remember, from Chapter 6, that gelatinisation is when the starch grains absorb moisture, swell up and burst, letting the polysaccharide called amylose leach out and trap the water into a gel; this thickens the sauce.)

Blended sauces that are heated show examples of conduction and convection currents. They need to be stirred all the time while they are being heated, so they do not go lumpy.

There are different types of blended white sauce.

White sauce with a roux base

This **roux** sauce involves mixing fat and flour (to form the roux base) and adding a liquid, which is often milk.

> **KEY TERMS**
>
> **Emulsifier:** An ingredient added to a mixture of a fat and a liquid that prevents the fat and liquid separating. An example is egg yolk in mayonnaise.
>
> **Infused sauce:** The milk, or liquid is cooked with an added ingredient, such as herbs or onion, before it is added to the other sauce ingredients.
>
> **Reduction sauce:** A sauce which is gently simmered to allow the liquid to evaporate, thickening the sauce.
>
> **Roux sauce:** a mixture of fat and flour which is heated before the liquid is slowly added to make the sauce.

White sauce: basic recipe

Ingredients
- 40 g flour
- 40 g butter or margarine
- 450 ml milk

Method
1. Add the butter or margarine to the saucepan and melt. Add the flour.
2. Stir the mixture over the heat until the mixture forms a ball. Allow to cook for two to three minutes, while you are stirring, to cook the flour.
3. Take the saucepan off the heat, and begin to add the milk very slowly, stirring well between each addition so no lumps are formed, and you have smooth liquid sauce after adding all the milk.
4. the saucepan back on a low heat and bring the sauce to the boil, stirring all the time. The sauce will thicken.
5. Keep stirring and simmer the sauce for two to three minutes to cook the flour completely.
6. You can then add any flavourings, such as grated cheese.

All-in-one sauce

An **all-in-one sauce** uses the same recipe as a roux sauce, but everything is put into the saucepan at the same time and the mixture is heated, stirring all the time, until it comes to the boil and thickens. This sauce will need to be cooked for about five minutes, stirring or whisking all the time, to cook the flour.

Infused sauces using a roux base

Examples of **infused sauces** are a velouté or a béchamel sauce.

A velouté is made with white stock but often when served with chicken it is 'finished' with cream.

Velouté: basic recipe

Ingredients

40 g flour

40 g butter or margarine

450 ml chicken, vegetable or fish stock

Method

Make the sauce the same way as the roux sauce, but use the stock instead of the milk.

Figure 11.27 **A velouté sauce**

A béchamel sauce is made the same way as a roux sauce, but the milk is gently heated with an onion, bay leaves and cloves or a pinch of mace, and left to take the flavours from these before it is used for the sauce.

Béchamel sauce: basic recipe

Ingredients

450 ml milk

1 bay leaf

10 black peppercorns

1 slice of onion

4 cloves or a pinch of powdered mace

40g butter

40g plain flour

Method

1 Place the milk in a saucepan with the onion, bay leaf, cloves or mace, and peppercorns. Place on the hob and bring to a simmer.

2 Turn off the heat, and leave the milk and flavourings to infuse for 30 minutes.

3 Strain the milk to remove the onion, peppercorns and cloves.

4 Make the sauce in the same way as a roux sauce above, using the infused milk.

Practical activity

1 When you are planning your practical sessions, see if you can use these recipes for sauces as part of your work. For example, you may choose to make a béchamel sauce for a lasagne instead of an all-in-one sauce, to add more flavour.

2 Plan and make a recipe using either a béchamel sauce or a velouté. Photograph your work and place it in your folder as evidence of demonstrating making one of these sauces.

Reduction sauces

Examples of reduction sauces are pasta sauce, curry sauce, gravy or meat sauce.

Gravy is often made using the concentrated juices that are left after a piece of meat has been roasted. A reduction means that the liquid is left to cook very slowly, while the water evaporates away, leaving a concentrated sauce behind that is full of flavour.

Meat reduction sauce

Ingredients

2 tablespoons of fat left from cooking your meat
1 chopped onion or shallot
250 ml beef or chicken stock
Salt and pepper

Method

1 Add the stock to the fat in the meat pan.
2 Add the chopped onion or shallot and put the pan onto the heat on the hob.
3 Keep stirring until the base of the pan is clean.
4 Reduce the heat and simmer, stirring occasionally, until the amount of stock is reduced by half, and the sauce has thickened.
5 Season with salt and pepper.

Practical activity

It is easy to make a gravy from the juices left when you have roasted or fried your chicken.

Use the recipe above to demonstrate this skill.

Meat juices, wine and balsamic vinegar can all be used as a basis for a reduction sauce.

Emulsion sauces

An **emulsion** is when fat, or oil and water are held together using an **emulsifier**. Oil and water usually will not mix and stay mixed. You can shake them together, and if you place the container down they will separate out again.

Sauces showing emulsification are mayonnaise, hollandaise and vinaigrette.

● In mayonnaise and hollandaise sauces, egg yolk is used as the emulsifier to hold the oil and vinegar together.

● In vinaigrette, mustard is often added to hold the oil and vinegar together. This sauce will need shaking before it is served.

Mayonnaise

Ingredients
- 300ml vegetable oil
- 2 egg yolks
- 1 heaped teaspoon powdered mustard
- 1 tablespoon vinegar
- Pinch of salt and pepper

Method
1. Place the egg yolks in a clean bowl.
2. Add the mustard and vinegar and whisk well.
3. Add a few drops of oil and whisk well.
4. Keep adding the oil, drop by drop, and whisking well each time. The sauce will emulsify and thicken.
5. Stop adding the oil when the desired thickness is reached.
6. Season with salt and pepper.

Hollandaise sauce

Hollandaise is a rich creamy sauce made from butter, egg yolks and lemon juice or vinegar; it is often served with vegetables, asparagus or fish.

Ingredients
- 3 egg yolks
- 1 tablespoon of fresh lemon juice
- 100g butter
- Salt and pepper

Method
1. Melt the butter in a saucepan, put to one side.
2. Place the egg yolks into a bowl, and put the bowl over a saucepan of simmering water. Make sure the base of the bowl does not touch the water.
3. Add the lemon juice and whisk together.
4. Slowly add the melted butter, whisking after each addition.
5. Once all the butter has been added you should have a smooth, thickened sauce to serve immediately.
6. Season with salt and pepper.

Practical activity

Choose two recipes that use these sauces. Make the sauces and tick them off on your list of skills.
- You can use the mayonnaise as an accompaniment to a salad, or as the basis for a Marie Rose sauce for a seafood cocktail.
- The hollandaise sauce can be used for Eggs Benedict or Eggs Florentine as a breakfast dish.

Skill 14: Set a mixture by removal of heat (gelation)

Gelation means to solidify by freezing or chilling.

If a mixture is chilled or frozen, it will set. You can see this happening in a cheesecake recipe, or in blancmange or custard on a trifle. Ice cream solidifies as it is frozen.

Figure 11.28 Cheesecake showing gelation

KEY WORD

Gelation: solidifying a mixture by chilling or freezing.

Practical activity

Make a chilled cheesecake using the following recipe, or choose one of your own. The following recipe is for a cheesecake, which is then chilled.

Cheesecake

Ingredients

100 g digestive biscuits, crushed into crumbs
50 g demerara sugar
50 g melted butter

500 g cream cheese
100 g icing sugar
1 teaspoon vanilla extract
200 ml whipped double cream

Method

1   Mix the demerara sugar, biscuit crumbs and melted butter in a bowl.
2   Spoon the mixture into a 20 cm springform cake tin that has been lined with greaseproof paper.
3   Press the mixture down firmly and put in the fridge to set.
4   Beat the cream cheese, icing sugar and vanilla extract together.
5   Add the whipped cream and mix well.
6   Spoon the cheesecake mixture onto the biscuit base, smooth down.
7   Chill for two hours until set.
8   Decorate with fruit such as raspberries if you like.

Tick the skills you have used off your list.

Exam practice questions

1   Identify the **three** main fat-based methods of cooking using the hob and give an example of a food to cook using each of these methods.   [6 marks]

2   List the stages of making a roux sauce. Explain what is happening as the sauce thickens.   [6 marks]

3   Describe the difference between a white sauce with a roux base and an all-in-one sauce.   [2 marks]

Skill 15: Set a mixture by heating

In Chapter 6 we looked at the functions of eggs. One function was **coagulation**. This is when the protein in the egg is denatured by heating, and it solidifies.

This property is used when making a quiche, or making choux buns or éclairs.

Practical activity

Making éclairs shows several of the required skills. Follow the recipe below to make the éclairs, and then list all of the skills you have used from the list.

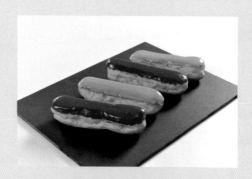

Figure 11.29 Chocolate éclairs

Chocolate éclairs

Ingredients

65 g plain flour
50 g butter
120 ml water
2 eggs
200 ml whipping cream
5 teaspoons icing sugar
100 g milk chocolate

Method

1 Preheat oven to 200°C, Gas Mark 6.
2 Grease a baking sheet.
3 Melt the butter and water in a saucepan. Increase the heat once the butter has melted.
4 Bring to the boil and immediately add the flour. Beat well over the heat until it becomes a smooth dough.
5 Cook for two minutes, stirring, until a glossy ball is formed.
6 Remove from the heat and leave to cool until the mixture feels slightly warm to touch.
7 Beat eggs in a bowl, then add a little at a time to the flour mixture, beating well until all the egg is combined and the mixture is smooth and glossy.
8 Either use a piping bag to pipe the mixture into éclair lengths, or use a teaspoon to make small round buns of mixture.
9 Bake in the oven for 15 minutes at 200°C, then turn down the temperature to 170°C for a further ten minutes until golden brown and crisp. **(Do not open the oven door during this time or the eclairs will collapse.)**
10 Remove the tray from the oven and make a small hole in the side of each éclair to allow the steam to escape. Return to the oven for five minutes to dry them out.
11 Remove from the oven, place on a cooling rack until cold.
12 Whip the cream until stiff and add the icing sugar.
13 Cut the éclairs down the side and pipe in the cream.
14 Melt the chocolate in a bowl over simmering water and use it to coat the top of the éclairs.
15 Leave the chocolate to dry and serve the éclairs.

Skill 16: Use of raising agents

Many recipes require the product to rise when cooked, to form a lighter result.

Mixtures can rise by:

- **incorporating air**, which will expand when heated and become trapped in a mixture when the gluten sets, such as in a creamed sponge mixture or a whisked meringue mixture
- **using a chemical raising agent**, such as self-raising flour, baking powder, bicarbonate of soda or yeast
- **using steam** in a mixture, such as choux buns or chocolate éclairs

Skill 17: Make a dough

The skills required for this can be covered by making shortcrust pastry or shortbread, bread and homemade pasta dough.

Shortening was discussed in Chapter 6. It is the ability of fat to coat the flour particles, forming a waterproof layer and preventing the gluten forming long strands, so the final product has a crumbly, 'short' texture. This happens when you create a 'rubbed-in' mixture for shortcrust pastry or shortbread.

Gluten formation happens during the kneading of bread dough. The protein called gluten in the flour is stretched during the kneading and forms long strands of protein which will act as the structure of the bread. This hardens as the bread is cooked, and traps the air to form the final product.

Fermentation is when the yeast in bread dough begins to grow, feeding on the sugar, and carbon dioxide is released causing the bread to rise.

Homemade pasta dough

Most people buy ready-made dried or fresh pasta. This is quicker and easier to use. However, you need to be able to show that you can make your own pasta.

Figure 11.30 Bread dough rising showing fermentation

Figure 11.31 Ready-made dried and fresh pasta

Look through your recipes and identify when you have used all these raising agents. Tick them off on your checklist, making a note of the recipes.

This is your evidence.

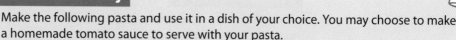

Practical activity

Make the following pasta and use it in a dish of your choice. You may choose to make a homemade tomato sauce to serve with your pasta.

Fresh pasta dough

Ingredients

300 g plain flour or Italian '00' flour
3 eggs
quarter of a teaspoon of salt

Method

1 Combine the flour and salt in a bowl.
2 Break the eggs into a small bowl and beat with a fork until they are mixed well.
3 Make a well in the centre of the flour and pour in the eggs.
4 Bring the mixture together using your fingers until a soft dough is formed. Add a little more flour if it is too sticky.
5 Knead well until the dough is no longer sticky, and is elastic to the touch.
6 Wrap in cling film and put in the fridge for 30 minutes to rest.
7 While the pasta is resting you can make the sauce of your choice.
8 You can then use a pasta machine to roll out the dough until it is very thin, or you can use a rolling pin.
9 Cut the thin, rolled out dough into the shapes you want.
10 This pasta will take four to five minutes to cook in boiling water.
11 Drain your pasta and top with the sauce.

Photograph your completed dish and evaluate using sensory descriptors.

Presenting a selection of recipes

Skill 18: Shaping a dough

Rolling out pastry, using a pasta machine, lining a flan ring, proving/resting

- If you have made a quiche, jam tarts or sausage rolls you have already demonstrated rolling out pastry.
- You will have lined a flan ring or case when making your quiche.
- You could have used a pasta machine when you were making the recipe above.
- You will also have allowed bread to prove and your pasta dough to rest.

Creating layers

This is a technique where you roll and fold pastry, such as puff pastry, to create layers. As the pastry cooks, the air that is trapped between each layer will rise, creating a flaky, light product.

Figure 11.32 Using a pasta machine

Make the following recipe for palmiers.

First you will have to make the puff pastry.

Puff pastry

Ingredients
250 g plain flour
quarter of a teaspoon of salt
250 g butter, at room temperature, but not soft.
About 150 ml cold water.

Method
1. Sieve the flour into a bowl with the salt.
2. Cut the butter into chunks, put into the bowl and then rub the butter into the flour.
3. Make a well in the bowl and add about two thirds of the water. Mix until you have a firm dough, adding more water if you need to. Cover with cling film and rest in the fridge for 20 minutes.
4. After 20 minutes, take the pastry and place on a lightly floured work surface. Knead gently and form into a rectangle. Roll the dough in one direction only, until the rectangle is about 50 cm long by 20 cm wide.
5. Fold the top third down to the centre, and the bottom third up and over that.
6. Give the dough a quarter turn and roll out again to a rectangle 50 cm long by 20 cm wide. Fold the bottom third up and the top third down as before.

Figure 11.33 Folding for puff pastry

7. Cover with cling film and rest the dough in the fridge for another 20 minutes.
8. After 20 minutes, repeat the steps above. You now have created several layers in your pastry.
9. The pastry can be stored in the fridge, wrapped in cling film for up to two days, or it can be frozen until you need it.

Palmiers

Ingredients
Puff pastry that you have made above
50 g caster sugar
2 teaspoons of ground cinnamon.

Method
1. Preheat the oven to 220°C, Gas Mark 7.
2. Roll out your puff pastry into a rectangle 50 cm by 20 cm.
3. Sprinkle the caster sugar and cinnamon all over the surface of the pastry.

Figure 11.34 Rolling up palmiers

4 Now start to roll up one side of the pastry, from the long side, very tightly until you reach the middle of the rectangle.
5 Then roll up the other side of the pastry until you reach the middle.
6 Cut into sections.
7 Place on a lightly greased baking tray and place in the oven for ten minutes until golden brown and well risen.

Figure 11.35 Cooked palmiers biscuits

Glazing and finishing

Adding a glaze, such as an egg wash, creates an attractive golden brown, glossy finish to a product, such as bread rolls or sausage rolls.

You could have **piped** the chocolate éclairs you made in skill 15 above.

You will have made the pasta recipe, and served it up with your homemade sauce. You can also use the pasta to make ravioli, which would involve shaping the dough.

Making flatbreads

Making **flatbreads** involves shaping your bread dough. Examples of flatbreads are naan bread, tortillas and chapattis.

Flatbreads are usually shaped after the bread has been left for the first rise. They are fried or baked after being shaped.

Figure 11.36 Glazed sausage rolls

Practical activity

Here is a recipe for naan bread.
Make the bread to accompany a curry that you have made.

Naan bread

Ingredients
250 g strong bread flour
1 teaspoon of dried yeast
1 teaspoon of sugar
150 ml warm water
1 tablespoon of oil

Figure 11.37 Naan bread

Practical activity

Find a recipe for stuffed ravioli, make your homemade pasta and create a ravioli dish with it.

Method
1 Mix all of the ingredients in a bowl, and work with your hands until a dough is formed, adding a little more water if needed.
2 Knead the dough for five minutes until a smooth, elastic dough is formed.
3 Cover the bowl and leave to rise in a warm place for an hour.
4 Divide the dough into six equal portions, roll each portion out to a round circle about 1 cm thick.
5 Fry the naan in a hot frying pan with a small amount of butter until they are brown and lightly risen.

Figure 11.38 Pinwheel biscuits

Making pinwheels

Pinwheels can be made with puff pastry or biscuit dough. You sprinkle the surface of the dough with your choice of filling and roll them up, using the same technique you used for the palmiers above.

Pinwheel biscuits are made by making two different flavours of biscuit dough, laying one on top of the other and rolling them up. You then slice down to form individual biscuits which you bake.

Practical activity

1 Find a recipe for pinwheel biscuits and make them.
2 Photograph your finished biscuits and place it in your folder.

Making pizza and calzone

Making homemade pizza and calzone uses bread dough for the base.

Figure 11.39 Pizza

● **Pizza:** the bread dough is rolled out into a circle after the first rise, and the topping is placed on the surface of the dough.
● **Calzone:** a calzone is a folded pizza, made into a pasty shape.

Practical activity

1 Choose a pizza recipe.
2 Divide your pizza dough into two equal-sized pieces. Use one piece to make as a regular pizza and the second piece to make a calzone. Compare the results.
3 Decide which you prefer.

Skill 19: Test for readiness

It is sometimes difficult to know if a product is cooked. It may look as if it is ready, but there are various techniques that you need to be able to use to see if a product is completely and safely cooked.

Using a temperature probe

Meat is a high-risk food. This means that if it is not cooked properly, it can cause food poisoning.

Figure 11.40 Calzone

One way to make sure that a dish containing meat is cooked is to use a temperature probe to make sure the centre of the meat, or dish containing meat, has reached a safe temperature to kill all the harmful bacteria.

The inside of chicken, beef, lamb or pork, or any dish containing these should reach 74–75°C to ensure all harmful bacteria has been killed.

Using a knife or skewer

You can check if a cake is cooked by inserting a knife or skewer and seeing if it comes out clean, with no mixture stuck to it. This means the cake is cooked.

Figure 11.41 Using a temperature probe to check chicken

If you stick a knife or skewer into the thickest part of a chicken leg, and the juice that runs out is clear, with no blood, it means the chicken is cooked.

Vegetables will be soft when a knife is stuck into them.

Using a finger or 'poke' test

This is often used when you are cooking a cake to see if the cake springs back when lightly touched. It is also used when kneading bread dough to see if it is elastic and springs back.

Using a 'bite' test

This is usually done with cooking pasta, to see if the pasta is cooked properly.

Using a visual colour check or sound

Often you can tell from the colour of a product if it is cooked. Biscuits, cakes and pastry products will turn golden brown when cooked.

A sound check can be used to see if a bread roll or loaf is baked, as it will sound hollow when the base is tapped.

Skill 20: Judge and manipulate sensory properties

How to taste and season during the cooking process

If you listen to or read anything by the top chefs in the world, they will all talk about the importance of tasting your recipes during the cooking process to make sure that the seasoning is correct.

You will need to demonstrate that you can do this. It is easy to get into the habit of doing this during all of your practical sessions. Remember to taste with a metal spoon that is easy to clean and NOT a wooden spoon.

Change the taste and aroma through the use of infusions, herbs and spices, paste, jus and reductions

The recipes that you select for your practical sessions will use different ingredients to add flavours, aromas and colours to your finished products.

An **infusion** is the flavour that is extracted from any ingredient by steeping or soaking them in oil, vinegar or water. Making a cup of tea is an example of an infusion.

Infused oils, such as herb flavoured oils, can be used in your recipes.

Herbs and spices will often be used in recipes.

Pastes such as Thai green curry paste can be bought at supermarkets to use in recipes.

Figure 11.42 Herb-infused oils

A **jus** is the juice that is obtained from the cooking of an ingredient, often meat. This is made into a thin gravy or sauce and served with the meat.

A **reduction** sauce was described in Skill 13. This is when the water is allowed to evaporate from a sauce, concentrating the flavours. This can be used in making a gravy for a meat dish.

Change the texture and flavour using browning (dextrinisation) and glazing; add crust, crisp and crumbs

Dextrinisation has been discussed in Chapter 6. It is when the starch on the surface of a product is changed into sugar by heat, and turns brown.

You may sprinkle sugar onto the top of an apple pie to make an attractive brown topping to the pastry.

Brushing a pastry dish with egg before cooking will create a glossy, golden brown texture to the surface, making a more attractive dish. This is called **glazing**. Look at Figure 11.36 showing glazed sausage rolls.

Examples of how you can add crust, crisp and crumbs include:

- Coating a food, such as fish, meat products like rissoles, or vegetables, in flour, egg and breadcrumbs before frying or baking to add a crunchy, crisp texture to the final product. Other coatings that can be used to create a crunchy finish are fine matzo meal, fine cornmeal, polenta or semolina, and Panko, which are flaky, Japanese breadcrumbs.
- Coating fish in batter before deep-frying to create a crunchy crust to the fish.

You may have demonstrated some of these techniques in Skill 3 'Prepare, combine and shape' and Skill 10 'Dry heat and fat-based methods using the hob'.

Figure 11.43 An apple pie showing dextrinisation   Figure 11.44 Panko breadcrumbs

---

Activity

Find some recipes using unusual coatings for meat, fish and vegetables. Add them to your folder of saved recipes until you get an opportunity to try them out during practical sessions.

KEY WORDS

Infusion: the flavours extracted from any ingredient by soaking or steeping in water or oil.

Jus: the liquid obtained from the cooking of an ingredient, often meat.

Reduction: when the water is allowed to evaporate from a sauce or liquid, concentrating the flavours.

Presentation and food styling

The more attractive food looks, the more likely people are to eat it. We all 'eat with our eyes' before we taste the food.

A **garnish** is an item or substance used as a decoration or embellishment accompanying a food dish, to give added colour or flavour. Garnishes are usually edible. Here are some ideas for garnishes.

Fruit garnishes

Citrus fruit flower: slice the peel of a lemon, a lime or an orange thinly and roll the peel into a flower.

Strawberry flower: place the strawberry on a cutting board with the pointed end facing up. Make several equal vertical cuts from the tip, without cutting through the strawberry and carefully fan out the thin cuts so it looks like a flower.

Strawberry fan: gently slice through the flesh of the strawberry and fan out.

Figure 11.45 Citrus fruit flower

Figure 11.46 Strawberry fan

Vegetable garnishes

Carrot flower: place a carrot slice on cutting board with largest cut end facing up. Using a sharp vegetable knife, cut down vertically about two-thirds of the way, do not cut completely through. Carefully carve each division into rounded petals and then cut out centres of the carrot slice to resemble a flower. You may have to cut a small amount off the bottom of the carved carrot at a slight angle so that it will sit on the plate.

Cucumber curl: this is a simple but very attractive garnish. Take a slice of cucumber about 0.5 cm thick. Cut a slit in the cucumber slice from the peel to the centre. Using your fingers twist the cut ends of the cucumber slice in opposite directions creating a 'curl'.

Figure 11.47 Carrot flower

Potato rose: these have to be made in advance, but you could make them at home the day before your practical session and bring them in.

Preheat the oven to 180°C. Slice the peeled potato into 0.25 cm thick slices. You can use a mandolin if you have one, or a very sharp knife. Put the slices in boiling water for one minute to blanch them, and then plunge them into ice water to stop them cooking.

Roll up the slices, overlapping each slice, to form the flower. Secure each flower with cocktail sticks, brush the tops with melted butter and bake in the oven for 45 minutes.

Cherry tomato flower: put the cherry tomato onto a chopping board, and using a vegetable knife, cut down the centre in a cross, not cutting right through the tomato. You now have four petals. Carefully remove the seeds and lay on the plate.

Tomato rose: peel the skin carefully from a tomato and roll up like a rose.

These are just a few ideas for garnishing. You can find many more ideas on the internet, with tutorials on how to make them.

Figure 11.48 Potato rose

Figure 11.49 Tomato rose

Other decorative techniques

- **Piping:** you can pipe icing or cream onto cakes or desserts to finish them attractively.
- **Glazing:** you can glaze dishes before they are cooked to give a shiny, golden brown finish, or you can glaze food after cooking, for example with melted butter to give a shine.
  - You can use jelly or aspic to coat savoury foods to give a clear, shiny coating.
  - Apricot jam is sometimes used to give a shiny coating to desserts and cakes.

Think about how to present your dish as attractively as possible.

WJEC EDUQAS GCSE Food Preparation and Nutrition

Working safely

During all practical sessions it is vitally important that you work in a way that will prevent any harm to yourself, and to the people eating your completed dish.

This means you have to be aware of anything that can cause harm during the cooking process.

Personal hygiene

You must follow the basic hygiene rules when preparing and cooking food:

1 **Tie long hair back:** this prevents your hair falling into food and contaminating it.
2 **Wash your hands:** always wash your hands before you start cooking, after handling raw meat and fish, after using the toilet and after blowing your nose or sneezing on your hands. Raw meat and raw fish are high-risk foods and contain harmful bacteria. Bacteria from the raw meat or fish can get onto your hands and be spread onto other foods or work surfaces. This is called **cross-contamination**. Washing your hands before you touch other foods will reduce the risk of this happening.
   Wash your hands after you have used any cleaning agents to wipe down surfaces as the chemicals may be harmful if they get into food.
3 **Remove any jewellery:** rings on your fingers can trap bacteria underneath. All rings should be removed before cooking.
4 **Wear a protective, clean apron:** your clothes have bacteria on them. You need to protect the food from these bacteria.
5 **Keep your nails short and clean, with no nail varnish:** nails can trap bacteria; flakes of nail varnish can fall into food, contaminating it.
6 **Cover any cuts or boils with blue kitchen plasters:** cuts and boils can contain harmful bacteria. Blue plasters are used because if they fall into the food they can be easily spotted.
7 **Do not cough or sneeze over food:** this will contaminate the food with bacteria.
8 **Do not cook if you are unwell, particularly if you have a stomach upset:** the bacteria causing the illness could be transferred into the food and spread to other people who eat the food.

Figure 11.50 Washing hands thoroughly reduces the risk of cross-contamination

Storage of foods and preventing cross-contamination during cooking

In Chapter 7, the subject of food spoilage is discussed in more detail. Look back at Chapter 7 to see how to store foods correctly.

Using equipment safely

All kitchens can be dangerous places. Knives, electrical equipment, hot ovens and hobs, spillages, and cleaning equipment and chemicals can all potentially cause harm.

Using knives

Earlier in this chapter we looked at different types of knives for each function. We also looked at using the bridge hold and claw grip for cutting vegetables.

- Make sure you are using the correct knife and grip when preparing food.
- Use different knives for raw and cooked food to avoid cross-contamination.
- Blunt knives are more dangerous than sharp knives, as they need more pressure and can slip and cut.
- Do not put knives into a washing up bowl and leave them, as someone can easily put their hand in the water and cut themselves.
- Keep knife handles clean and grease free so they do not slip when using them.
- Never walk around the kitchen holding a knife with the blade pointing outwards.
- After washing and drying the knife, return the knife to the correct storage place.

Using electrical equipment

- Check all machinery is in good working order, with no frayed cables or loose plugs.
- Do not handle electrical equipment with wet hands or you could get an electric shock.
- Never drape electric cables in or over water or wet patches on work surfaces.
- Only one person uses the equipment at a time.
- Keep your hands away from the moving parts.
- Wash blades and beaters, dry carefully and replace onto the equipment.
- Place the equipment on a work surface so it is stable, and not liable to topple or fall.
- Turn all equipment off and unplug when you are not using it.

Hot ovens, hobs and equipment

- Use oven gloves or oven cloths, not tea towels, to place food in and remove food from the oven to prevent getting burnt.
- Turn pan handles away from the other hot plates if they are being used, so the handles do not get hot and burn you.
- Take care when moving hot pans or dishes; ask people to move out of the way if you are sharing a cooker or hob.
- Never put hot dishes directly onto a work surface or chopping board; use a cooling rack or trivet.
- Do not leave oven doors open or hot plates turned on or lit when you are not using them. Do not leave a grill unattended.
- Do not clean the oven or hob while it is still hot.

Spillages

- Mop up any spillages on the floor or work surfaces immediately to prevent people slipping or burning with hot liquids.
- Use the appropriate floor or dish cloths to prevent cross-contamination.

Cleaning equipment

- Clean up as you go while cooking so that there are not piles of dirty dishes over your work surface or in the sink area. Wash up dirty pots and pans and dry and put away in the correct place.

- Wipe surfaces down regularly to prevent contamination.
- Use the appropriate cloths and cleaning fluids for each area. Do not mix cleaning fluids.
- Ensure food wrappings and peelings are cleared into the correct bin before you start preparing food.
- Keep bins covered. If you have to touch the bin to remove the lid, make sure you wash your hands before preparing your food.

KEY POINTS: WORKING SAFELY

- Follow the personal hygiene rules that ensure you are protecting both yourself and your customers, or the people eating your prepared food – they will keep you safe and prevent cross-contamination of foods.
- Ensure the food you are using has been correctly stored and is in date.
- Make sure that you take care when using sharp knives and equipment.
- Check that all electrical equipment is safe before you use it.
- Protect yourself while dealing with hot surfaces and removing or placing food in the oven by using oven cloths.
- Clean up spillages, and correctly clean and wash all equipment before you put it away.

Activity

There are lots of food safety and hygiene quizzes and activities on the internet.

Visit the Food Standards Agency website and complete the following quizzes:

- **Food Hygiene quiz:** www.food.gov.uk/ sites/default/files/ multimedia/pdfs/ hygienequiz.pdf
- **Food safety: what do you know?** www. food.gov.uk/news-updates/campaigns/ germwatch/fsq

Exam practice questions

1 Identify **three** personal hygiene rules that should be followed when cooking. [3marks]

2 Explain what cross-contamination is, and discuss how it can be prevented while cooking high-risk foods. [6 marks]

3 Discuss why someone who has a stomach upset should avoid preparing food. [4 marks]

Working independently

As you become more confident in your practical work, you will be expected to work independently. You can only gain the higher marks in practical sessions if you do not ask for help all the time.

To be able to work independently, you need to:
- have all the skills necessary to carry out your chosen task
- be well organised, by having a detailed time plan and have all the equipment and ingredients ready
- clear up and wash up as you go along
- be hygienic in your working and food preparation.

Using sensory descriptors appropriately and correctly

Activity

Each time you cook a dish, use words from the table above to write a full description of your dish.

When you make your chosen recipes in practical sessions, or for your final assessments, the food will be photographed. The examiner will not be able to look at or taste the dishes, but will need to rely on your description of the food to give you a mark. It is therefore important that you get into the habit of describing how your food looks, smells and tastes.

The sensory descriptors below give you all the words you should be using to describe your food.

Describing taste, flavour and smell

- Acid
- Aftertaste
- Astringent
- Bitter
- Bland
- Burnt
- Creamy
- Dry
- Hot
- Fatty
- Metallic
- Old
- Piquant
- Pungent
- Salty
- Sharp
- Sickly
- Soggy
- Sour
- Spicy
- Tangy
- Tart
- Zesty

Describing appearance

- Appetising
- Attractive
- Bubbly
- Cellular
- Clear
- Cloudy
- Coarse
- Colourful
- Colourless
- Crumbly
- Dry
- Fat
- Flat
- Foamy
- Fresh
- Greasy
- Grained
- Healthy

Describing texture

- Adhesive
- Bouncy
- Brittle
- Bubbly
- Cellular
- Chewy
- Close consistency
- Coarse
- Cohesive
- Cold
- Crisp
- Crumbly
- Crunchy
- Crystalline
- Dry
- Elastic
- Effervescent
- Fibrous
- Firm
- Fizzy
- Flabby
- Flaky
- Foamy
- Juicy
- Lumpy
- Moist
- Mushy
- Pulpy
- Powdery
- Slimy
- Smooth
- Soft
- Soggy
- Spongy
- Springy
- Sticky
- Stiff
- Stretchy
- Stringy
- Syrupy
- Tacky
- Tender
- Thick
- Thin

Exam practice questions

1 Identify **three** reasons why it is important to make a detailed time plan for your practical work [3 marks]
2 List **four** descriptive words that you could use to describe how a Victoria sandwich cake looks when it has been removed from the oven before it is decorated. [4 marks]
3 Discuss why it is important to use descriptive words in your Non-examination assessment work. [4 marks]

CHAPTER 12
Developing recipes and meals

Learning objectives

In this chapter you will learn:

- how to consider the influence of lifestyle and consumer choice when developing meals and recipes, and how to make adaptations to recipes to address current dietary advice and lifestyle patterns
- how to consider nutritional needs and food choices when selecting recipes
- how to develop and review your recipes and make improvements to them
- how to manage your time and the costs of recipes
- how to use your testing and sensory evaluations to improve or adjust your recipes
- how to explain and justify your choices and how to present your ideas about your chosen recipes and cooking methods
- how to make decisions on which techniques are appropriate to use, using your understanding of nutrition, culinary traditions and food preparation.

Considering the influence of lifestyle and consumer choice when developing meals and recipes

In Chapter 4 Plan balanced diets we looked at specific religious beliefs, ethical beliefs, illnesses, allergies and dietary conditions that influence what people chose to eat, or can eat.

When selecting recipes for a particular task, you may have to consider any one of these criteria; or you may have to develop or alter a recipe because a member of your family suffers from a dietary condition.

Figure 12.1 A Halal Butcher's shop

Religious beliefs

To recap, certain religions have restrictions on what they can eat, or they can only use food that has been slaughtered in a certain way (e.g. Muslims have Halal foods and Jews have Kosher foods). Catering for one of these religions means you would have to find a specific butcher who could sell Halal meat, or Kosher meat.

Ethical choices

Ethical choices include those who follow a vegetarian diet, including ovo-lacto vegetarians, lacto vegetarians and vegans. These have different requirements:

- Ovo-lacto vegetarians eat eggs and dairy produce but no meat or fish.
- Lacto vegetarians eat dairy products but no eggs, meat or fish.
- Vegans eat no animal products at all (including no dairy products, eggs, meat or fish).

The table below shows how a recipe can be altered to cater for each of these dietary categories.

Recipe and food item for a non-vegetarian Cottage Pie	Ovo-lacto vegetarian	Lacto vegetarian	Vegan
250 g beef mince	250 g Quorn™ mince	200 g Textured Vegetable Protein	200 g Textured Vegetable Protein
1 onion	1 onion	1 onion	1 onion
Beef stock cube	Vegetable stock cube	Vegetable stock cube	Vegetable stock cube
Herbs, salt and pepper	Herbs, salt and pepper	Herbs, salt and pepper	Herbs, salt and pepper
150 g carrots	150 g carrots	150 g carrots	150 g carrots
1 tin chopped tomatoes	1 tin chopped tomatoes	1 tin chopped tomatoes	1 tin chopped tomatoes
350 g potatoes	350 g potatoes	350 g potatoes	350 g potatoes
25 g butter	25 g butter	25 g butter	25 g non-dairy spread
25 ml milk	25 ml milk	25 ml milk	25 ml soya milk

Table 12.1 Alterations to a recipe for different types of vegetarians

It was quite simple to change this recipe for cottage pie to suit three different diets.

Activity

Look at the alterations to the recipe in Table 12.1 and answer the following questions:
1 Identify all the alterations in the recipe.
2 Why can the ovo-lacto vegetarian eat Quorn™, but the lacto vegetarian cannot?
3 Why has the butter been substituted for non-dairy spread for the vegan?

Some recipes will be more difficult, as ingredients have to be checked carefully, especially if you are using processed foods for a recipe.

Consumer choice can mean that a particular ingredient or type of ingredient is used. Some people prefer to use **organic** foods where possible; others like to buy **Fairtrade** products or local produce.

Figure 12.2 Soil Association organic logo

Stretch & challenge

1 Find out how a producer has to alter the way they farm or produce a product to allow it to be identified as **organic**.

2 Make a presentation for the rest of your class to show what has to be done, and how long it takes for a farm to be recognised as producing organic products.

Pregnancy

Pregnancy is not an illness, but as we can see in Chapter 4, there are a lot of foods that a pregnant woman needs to avoid. It is important to check recipes when cooking for pregnant women.

Illnesses

Other illnesses include Type 2 diabetes, cardiovascular disease, coronary heart disease and coeliac disease. All these conditions require specific alterations to a diet, and often the substitution of alternative ingredients to cater for the illness.

Type 2 diabetes is an increasing problem in the UK. There are currently over 3 million sufferers with diabetes, and 95 per cent of these have Type 2 diabetes.

No food is 'out of bounds' for a diabetes sufferer, but food choices are important. Eating a balanced diet is important, as well as eating regular meals to maintain a steady blood sugar level. Some Type 2 diabetics can control their condition simply by eating the right foods.

Activity

1 The website **www.diabetes.co.uk** has a large number of tips and recipes specifically for this condition.

Look at the website and create a day's menu for someone who is suffering from Type 2 diabetes.

2 Make sure you are providing a balanced diet which contains all the nutrients needed to keep healthy.

Cardiovascular disease and **Coronary heart disease** are related conditions. A healthy, balanced diet is the best way to lower or maintain lower cholesterol conditions, reducing the risk of heart attack or stroke.

Practical activity

1. The British Heart Foundation website (**www.bhf.org.uk**) has many tips and recipes for sufferers of these diseases.

   Use the above website to find a recipe for a main meal which is suitable for someone suffering with one of these conditions.

2. Plan and cook the recipe, and evaluate the result using sensory descriptors.

Sufferers of **coeliac disease** have to avoid any foods containing gluten, so using special products which are gluten-free as substitutes will make the food suitable. However, if you decide to use gluten-free flour, you need to remember that it has different properties to ordinary flour, and will not support a structure, as in bread or cakes made with ordinary flour. There are special recipes for gluten-free bread and cakes which use other ingredients to help provide a structure to support the risen dough.

Ordinary Wholemeal Bread	Gluten-free brown bread
500 g Wholemeal Bread Flour	325 ml semi-skimmed milk
1 sachet dried yeast	2 large eggs
1 teaspoon of sugar	1 teaspoon white wine vinegar
Pinch of salt	450 g gluten-free brown bread flour
325 ml warm water	2 tablespoons caster sugar
1 tablespoon vegetable oil	1 sachet dried yeast
	3 tablespoons olive oil
	Pinch of salt
Method	**Method**
1 Mix the flour, salt, sugar and yeast in a bowl.	1 Slightly warm the milk.
2 Add the water and mix into a dough.	2 Crack the eggs into a large bowl, add the vinegar and the warmed milk and stir.
3 Add the oil and knead for ten minutes until the dough is soft, pliable and elastic.	3 Place the flour, salt, sugar and yeast in a bowl and mix.
4 Place back in the bowl, cover with a tea towel and leave in a warm place until it has doubled in size.	4 Add the milk mixture and combine until a sticky dough is formed.
5 Turn the dough out onto a floured board and knead again until smooth and elastic.	5 Add the olive oil and bring it together with your hands into a ball.
6 Shape the dough and put onto an oiled baking sheet or into a greased loaf tin.	6 Place on a lightly oiled baking tray and cover with a damp tea towel.
7 Cover with a tea towel and leave to rise in a warm place for 35 minutes.	7 Leave to rise in a warm place for 1 hour, or until doubled in size.
8 Preheat the oven to 200 degrees Celsius, Gas Mark 6.	8 Preheat the oven to 200 degrees Celsius, Gas Mark 6.
9 When the dough has risen, bake the loaf for 35 to 40 minutes until golden brown and cooked through. The base should sound hollow when tapped lightly.	9 When the dough has risen, place the bread into the oven for 35 minutes, or until golden brown and cooked through.

Table 12.2 Ordinary wholemeal bread recipe and gluten-free brown bread recipe.

Figure 12.3a–b Ordinary wholemeal bread and gluten-free brown bread

Practical activity

1 Work with a partner. One person makes the ordinary bread and the second person makes the gluten-free bread.
2 Compare the outcomes. Ask a panel of tasters to use sensory descriptors to describe the two breads.
3 Do you think that the gluten-free bread is an acceptable alternative for a coeliac sufferer?

Family lifestyle patterns

● Many families have two working parents, or are single-parent families with one working parent.
● Some families have a reduced budget, requiring care to be taken when planning meals due to the amount of money they have to spend.
● Some families rarely sit down to eat together, as members of the family have different interests or social commitments at different times. There is also less emphasis on eating together as a family nowadays.

Working parents may have limited time to spend cooking meals. It is very easy for them to begin to rely on ready-made processed food that just has to be heated up when they arrive home from work, tired and with little time to prepare healthy, home-cooked meals.

However, it is possible to make food from fresh ingredients, even if you do work full time.
● You can batch cook at the weekend, and freeze portions of food that can be reheated on a week night and served with freshly cooked vegetables. Examples of foods that are easy to cook in large quantities are homemade pasta sauces, lasagnes, chillis, cottage pie, soups, stews, meatballs, fish cakes, fish pie, pasties, cannelloni, meatloaf and homemade beef burgers. There are many websites that offer recipes that can be frozen:
  ○ **www.bbcgoodfood.com/recipes/collection/freezable**
  ○ **www.bbcgoodfood.com/recipes/collection/batch-cooking**
  ○ **www.foodnetwork.co.uk**, then search for Batch Cooking Recipes.
● Many websites also offer 20- or 30-minute recipes, such as Jamie Oliver's 30-Minute Meals recipes (**www.jamieoliver.com**).
● Supermarkets often have recipe cards or magazines which include quick recipes.

Families on a limited budget have to look carefully at the amount of money they can spend on food.

There are many ways to save money, and still provide a healthy, balanced diet:

1 Look for products that are on offer in a supermarket. Plan a meal around this offer.

2 Buy products such as meat, fish and vegetables from the 'reduced' section, where they are nearing their sell-by or use-by date. Either use them that day, or freeze the meat and fish portions for later in the week. Use the fruit or vegetables that night, or make soup, sauces or desserts from them for the next day.

3 Buy own-brand products rather than premium brands. The difference in price can be great, but the quality is not usually very different.

4 Shop at cheaper supermarkets such as Lidl and Aldi, where good quality products are available at much lower prices.

5 Use canned or frozen fruit and vegetables instead of fresh. They often have the same nutritional values but are less expensive.

6 Shop at markets rather than supermarkets as you can often buy smaller quantities of ingredients rather than a large pack.

7 Use leftovers to make another meal. A whole chicken can make a roast dinner, chicken curry the next day and either chicken soup or stew from the carcass for the third day.

8 There are sometimes 'Weigh and Save' shops which have loose, dried goods that you can buy in small quantities for recipes.

Families that do not often eat together can still have healthy meals if they freeze individual portions of food in the freezer. This means each person can reheat the food when needed.

Figure 12.4 A whole chicken can provide three different meals

Current dietary advice

In Chapter 4 we looked at current government advice for a healthy diet. A reminder of the eight suggested rules for this are:

1 Base your meals on starchy foods.
2 Eat a minimum of five portions of fruit and vegetables every day.
3 Eat at least two portions of fish a week, one of which should be oily fish.
4 Cut down on saturated fat and sugar.
5 Eat less salt, no more than 6 grams a day for adults.
6 Drink plenty of water.
7 Do not skip breakfast.
8 Get active and try to be a healthy weight.

There are also RDI (Recommended Daily Intake) amounts for all the main macronutrients and micronutrients.

To keep healthy, people need to take note of these rules when planning meals. A quick reference is to use the Eatwell Guide diagram.

Recipes can be altered and adapted to fit in with this advice.

Activity

Select two recipes – one for a main course and one for a dessert. Suggest adaptations and changes to make the recipes fit in with the above eight suggested rules for a healthy eating pattern.

KEY POINTS: CONSIDERING THE INFLUENCE OF LIFESTYLE AND CONSUMER CHOICE

- Religious and ethical beliefs will have an impact on the choice of ingredients for recipes.
- Illness and allergies will need to be taken into account when selecting ingredients for your chosen recipes.
- Family lifestyles, including working patterns and budget will have a major impact on choice of recipe and ingredients.
- Current dietary advice will need to be taken into consideration to plan a healthy, balanced diet.

Exam practice questions

1 Suggest **three** ways that a family on a limited budget could cut down on the amount of money they spend on food. [3 marks]

2 Explain why gluten-free flour will not produce a loaf which is the same as a loaf made of ordinary bread flour. [6 marks]

Considering nutritional needs and food choices when selecting recipes

In Chapter 4 we looked at the nutritional needs of different age groups, and discussed the amount of nutrients that each age group needs.

We also looked at people who had deficiencies in their diets, causing conditions such as **anaemia**, people with **intolerances** or **allergies**, and also people with **high energy needs.**

When selecting recipes, you will need to make decisions about the ingredients based on the requirement of the brief.

Intolerances

You will need to check ingredients carefully, and substitute any ingredients that could be dangerous or cause illness or discomfort. Check with the person or people that you are cooking for before you begin planning the dish as to whether they suffer from any allergies or intolerances.

- Remember: nut allergies can be life-threatening.
- Lactose intolerance will cause discomfort for the sufferer, and many lacto-free alternatives are available.
- Some people can be allergic to eggs or shellfish.

Lifestyle choices

If you are cooking for vegetarians, you will need to find out which type of vegetarian they are. In Chapter 4 we looked at the different types of vegetarianism.

There are many substitute ingredients that can be successfully used as alternatives in recipes to cater for these different groups. An example of how to adapt a recipe is shown in Table 12.1.

Cooking processes and methods

In Chapter 6 we looked at the effect of cooking on food, and how some nutrients are destroyed by cooking methods.

If you are trying to conserve nutrients you need to check how certain nutrients are destroyed by some cooking methods.

Table 6.2 shows the effect of cooking on the main macronutrients and micronutrients.

Table 6.3 shows which vitamins are destroyed by cooking methods.

It is important to minimise nutrient loss during cooking, so care must be taken to select the correct cooking method for your food.

You may also be cooking for elderly people who have difficulty chewing. Therefore, you need to make sure that you do not include very chewy or hard foods which they are unable to eat. Elderly people with arthritis may have difficulty gripping knives and forks, and cutting foods.

If you are cooking for young children, **portion control** is important. Young children do not need as much food as older children or adults, and small, easy-to-eat food is often more attractive to encourage children to eat.

Figure 12.5 A meal suitable for an elderly person who has difficulty chewing or cutting food

Practical activity

Plan and cook a main course suitable for a 5-year-old child.

1 Check back to Chapter 4 to see which foods they need to have, and what quantities are recommended as RDIs for that age group (see Table 4.1).
2 Explain why you have chosen that recipe, and identify where all the required nutrients are found.
3 Make the dish and photograph it.
4 Create a chart of sensory descriptive words, or make a 'smiley face' chart for the child to complete and, if possible, invite a 5-year-old child to eat it and evaluate it for you.

Figure 12.6 Children's portion sizes compared to adult portion sizes.

Portion control is important in catering for someone who is trying to **lose or maintain their weight.**

In Chapter 4 we discussed ways to help someone lose weight. Portion control is important, particularly if you are reducing calorie intake. Careful weighing of ingredients and calculations of calorie values, using Table 5.1, will mean that you are much more likely to provide a meal which has the exact calorie values to enable the person to lose or maintain weight. Fatty foods have a much higher calorific value than starchy foods.

The amount of calories is also important for someone with **high energy needs**. If the person you are cooking for has been ill, has a specific dietary condition that means they do not absorb energy from food so easily, or they are an athlete or sports person, you will need to calculate their **BMR** and **PAL** levels to find the amount of calories needed in their diet, and then use the calculations in Table 5.1 to work out how many calories are in the meal or dish you are planning to cook.

KEY POINTS: CONSIDERING NUTRITIONAL NEEDS AND FOOD CHOICES

- When selecting and planning individual dishes and meals, you will need to take nutritional needs of your target group into consideration.
- Lifestyle choices and food intolerances will need to be considered when selecting and making different dishes.
- The method of cooking will need to be considered in order to preserve nutrients in your chosen dish or meal.
- The age of the person you are cooking for will have an impact on the size of portion you are serving up.
- The amount of energy and calorie content of your dish or meal will need to be calculated if you are cooking for someone who is trying to lose or maintain their weight.
- Someone with high energy needs will need to consume more calories, and calculations of their requirements will have to be made before selecting an appropriate dish or meal.

Exam practice questions

1 Identify **two** dietary conditions when you would have to use alternative ingredients, and explain why these ingredients have to be substituted with an alternative. [4 marks]

2 Discuss reasons why elderly people may need a softer diet than a younger person, and make **two** suggestions for a main course and dessert for an elderly person. [6 marks]

Developing, reviewing and making improvements to your recipes

If you are in the habit of evaluating all your cooking, and are using sensory descriptors from Table 11.4 to make a full explanation of how your dish looks, smells and tastes, you will be able to suggest improvements for that dish or meal if you were going to cook it again.

You may also have used **tasting tests** to find out people's opinions of your foods. These can help you to suggest alterations and improvements. Tasting tests are discussed in Chapter 10 Factors affecting food choice.

A step before that is to be able to look at a chosen recipe and identify any changes or improvements that can be made during the planning process **before you make it**, to enable the dish to fit your brief.

The following points need to be looked at:

- **Are you using appropriate ingredients?** If the dish or meal is for a particular age, dietary condition, illness or intolerance, have you checked that all the ingredients are suitable for that person? If not, what changes do you need to make?

- **Are you using the correct process and cooking method?** Will the preparation and cooking processes preserve nutrients? Will the food discolour while it is waiting to cook? Is it going to be cooked in time? Can it be reheated safely? Will it look attractive?
- **Are the portion sizes correct?** Who is the food being cooked for? Have you calculated the energy content if the food is for a low-calorie diet or high-energy diet?

Managing your time and the cost of recipes

In Chapter 11, Tables 11.1 and 11.2 gave examples of time plans that could be prepared when you are in the planning process for your chosen dish or meal. This is important to see if the dish or dishes you have chosen to cook fit in the amount of time you have. It also shows your ability to manage the cooking of several dishes at the same time.

The cost of recipes is also important, not only if the brief requires a low-cost meal, but also to ensure your own family is not spending excessive amounts of money to provide ingredients for your dishes.

You need to be aware of the cost of ingredients, which can be expensive. Many meats, fish and cheeses are expensive, and it is possible to select less expensive alternatives.

Look at the list of suggestions on page 250 (Family lifestyle patterns) for ways for families on a limited budget to save money. Incorporate these ideas into your planning processes. Identify where you have substituted less expensive ingredients.

Using your testing and sensory evaluation skills

You need to be able to check when food is cooked, using the techniques discussed in Chapter 11, Skill 19, pages 236–237. This includes using a temperature probe, a knife or skewer, using a finger or the 'poke' test, using a 'bite ' test and using a visual colour check or sound check.

During the cooking process, you need to be tasting your food for the correct seasoning and adjusting the amount added to produce the most flavoursome outcome. This is discussed in Chapter 11, Skill 20, pages 237–238. Remember that it is easy to add more seasoning, but very difficult to correct the over-seasoning of food.

You may also need to adjust cooking times. For example, if you change the portion size of the recipe, by reducing the servings from four to two, as well as halving the quantities in the recipe, the cooking time may need to be reduced.

You might need to add more colour or texture to the dish. This can be achieved by adding colourful side dishes, accompaniments or a simple garnish.

You have to be able to present your food attractively, using garnishes in the most aesthetically pleasing way.

> **KEY POINTS:**
> - All the planning processes before you cook should take into account the use of appropriate ingredients for your brief.
> - Make sure you have a detailed time plan, and that you have looked at the cost of the ingredients, and substituted less expensive ones if necessary.
> - Ensure you are tasting and adjusting seasoning during cooking.
> - Remember to adjust the cooking times if you have reduced or increased the recipe size.
> - Present your dishes in the most attractive way by adjusting colour, texture or by adding garnishes.

Exam practice questions

1 Suggest **two** different seasonings you could add to flavour a dish to replace salt if you are planning a dish for a low-salt diet.   [2 marks]

2 Identify **one** way in which you could reduce the cost of a fish dish and explain how this would reduce the overall cost.   [2 marks]

3 Discuss the benefits of garnishing a food to make it look more attractive.   [4 marks]

Explaining, justifying and presenting your ideas

You need to be able to suggest alternative ingredients, processes, cooking methods and calculate portion sizes, and to be able to **explain** and **justify** why you have made these changes.

The examiner needs to know that you understand why you have made any changes. Your **justification** and **explanation** should include:

- Reference to any scientific basis for the change. An example of this could be:

> This recipe for a fennel tart has a filling made of crème fraiche and a velouté made from the fennel stock. It is then baked in the oven. I think the filling is going to be too liquid, and will not look attractive, as it will run out of the tart when the tart is cut. I am going to add a beaten egg to the velouté, as this will denature during the cooking and coagulate, causing the filling to set slightly, and make a more attractive outcome for presentation.

- Reference to any dietary need for the change. An example of this could be:

> I am making this dish for someone who is lactose intolerant. This means they lack the enzyme to digest milk sugar, or lactose. I am going to substitute the semi-skimmed milk for lacto-free milk.

- Reference to any taste testing or practice sessions you have carried out. An example of this could be:

> When I made my selection of pasta sauces I used a ratings test for my tasters to find the most popular one. This was the one with the added herbs, so I am selecting this as my final choice.
>
> During my practice sessions, I found that the pastry shrank during cooking as it was not given sufficient time to rest. I will allow an extra ten minutes in the fridge to make sure it cools down sufficiently.

- Reference to any change because of timings or cost. An example of this could be:

> My dish is too expensive for a family on a low budget, so I am going to use a less expensive white fish, called saithe or coley, instead of cod. This is half as expensive as cod.
>
> The dish I have chosen takes two hours to cook, and I only have one and a half hours, so I will be making the pastry at home and bringing it in ready-made to save time. I will also cut up the vegetables and grate the cheese at home. I have already shown the skill of making shortcrust pastry in a previous practical session, and have also shown the other skills previously.

- Reference to any extras ingredients, seasonings or additions of colour or texture that you have included. An example of this could be:

> During the practice sessions my tasters all said the food did not look attractive, as it was not very colourful. I am going to make some tomato flowers to go on the top, glaze it with egg yolk before it is cooked to give it a shiny, golden brown finish, and serve it with some steamed mangetout to add a contrast in colour.

KEY POINTS:
MAKING
DECISIONS
ABOUT
APPROPRIATE
TECHNIQUES

- You need to use prior knowledge to explain and justify your ideas and alterations to a recipe or meal.
- Keeping a record of your practical sessions and the evaluations will help you to become more confident, and be able to improve your cooking and food presentation.
- Ensure you are referring to nutrition, food science, ethical and religious beliefs and cooking processes whenever you make an alteration, using whichever is appropriate.

Making decisions about which techniques are appropriate to use during preparation and cooking

The examiner will need to see that you understand the need of when and how to alter the techniques you are using during food preparation and cooking.

These should be based on the following:

- Your understanding of nutrition.
  - Chapter 2 in this book explains the types, roles and functions of macronutrients and micronutrients. It also explained how these nutrients work in our bodies, and what happens if we eat too much, or not enough of these. Your planning will need to show that you have taken this information into account while deciding what to cook.
  - Chapters 3, 4 and 5 look at different energy requirements of individuals, planning balanced diets and calculating energy content and nutritional values of dishes and meals. You must use this information and refer to it during your planning process.
  - Chapter 6 look at how heat affects nutrients. You need to show that you can use all this knowledge when deciding how to cook and prepare your ingredients for your practical sessions. If you are trying to conserve nutrients, state which method of cooking does this, for example steaming vegetables instead of boiling to conserve vitamin C.
- **Your understanding of food.** Chapter 6 looks at the science of food, what happens when it is cooked and how the properties of different nutrients are used to create the dishes we cook. You need to show that you can alter the techniques or type of ingredient to fulfil the brief or recipe.
- **Your understanding of different culinary traditions.** You need to be able to alter a recipe or ingredient to cater for a different religion or culture which has sets of rules for the food that they can eat.
- **Your understanding of food preparation and cooking.** You will have tried out many recipes and techniques when making food. Each time, your evaluation should have included what went well, and worked for you, and what needed improvement. Keeping a note of these will help you to produce better outcomes, be more confident and be able to change ingredients or cooking processes to ensure a successful result is achieved. If you include information for the examiner about what you have done, and why you have done it, you are showing your understanding.

Exam practice questions

1. Write an explanation of what happens to the protein in an egg when it is fried. Use the correct scientific terminology. [4 marks]

2. Explain how you would alter the ingredients in a recipe to cater for someone who is lactose intolerant. Give examples of replacement ingredients. [4 marks]

3. Suggest **two** improvements you could make to a chicken pasta dish that has been criticised for not being colourful enough. [2 marks]

SECTION 7

Non-examination assessment: Food preparation and nutrition in action

This section includes the following chapters:

Chapter 13 Assessment 1: Food investigation assessment

Chapter 14 Assessment 2: Food preparation assessment

As part of your GCSE course, you will be required to produce two assessments: the Food investigation assessment; and the Food preparation assessment, which will show that you are able to research, plan, trial and justify choices as well as work safely and hygienically, producing excellent practical results. You will also need to show that you can evaluate your practical work.

CHAPTER 13
Assessment 1: Food investigation assessment

The first assessment will be a scientific investigation into the working characteristics and the functional and chemical properties of foods.

You will need to include evidence to show that you can:

- **interpret and investigate** your chosen assessment, create a plan of action and propose a hypothesis or make an educated guess of what you think the outcome will be
- **carry out your plan of action** with trialling, developing and testing
- **analyse** the data and results and draw conclusions
- **explain** whether your prediction was proven.

The required word count for your completed assessment project is 1500 to 2000 words; your completed project will need to be printed in **A4 paper** (or A3 equivalent), using **font sized 11 or 12**.

This assessment is 15 per cent of the total of your GCSE marks.
- You will be allowed up to **eight hours** to carry out this assessment.
- You will be able to gain a total of **30 marks** for this assessment.
- You will get a choice of **two** different assessment tasks.

The marking is divided into:
- **5 marks** for the interpretation and investigation.
- **15 marks** for the carrying out of the plan of action, the trialling, developing and testing.
- **10 marks** for the analysis and explanations.

This assessment will require you to look at the working characteristics, and functional and chemical properties of a given ingredient or ingredients.

You will be expected to work hygienically and safely at all times.

Section 4 of this book looks at and describes **The science of food**. You may find it useful to refer back to this section when preparing for your Food investigation assessment

An example of a Food investigation task could be as follows:

Figure 13.1 You will be asked to investigate the working characteristics and functional and chemical properties of a given ingredient or ingredients – for example flour

> **Flour is available in several different types.**
>
> **Investigate the working characteristics and the functional and chemical properties of *three* different flours, creating a product from each type of flour that shows the best use for that flour.**
>
> **This assessment must be supported by investigational work.**

Section A: Interpretation and investigation (5 marks)

- For the example task, you would start by looking at what flour is. You would then write a paragraph explaining what flour is made from, what it is used for, that it contains starch and any other relevant information.
- Then you'd look at all the different types of flour that you can buy. Examples of these are:
  - white self-raising, wholemeal self-raising
  - white plain, wholemeal plain
  - strong white bread flour, wholemeal bread flour
  - italian '00' flour
  - gluten-free flour.
- You would then investigate what the properties of these flours are and what they are used for, then write up what you find. You will need to include some scientific descriptions and use the correct terminology to describe what happens when these flours are used.

  An example of this could be:

> Self-raising flour contains raising agents, which will produce carbon dioxide gas when the flour is mixed with a liquid. As the product is cooked, the gas bubbles expand and cause the product to rise. This can be seen when you are making a cake. Self-raising flour is soft flour, which is made from spring wheat. It does not contain much gluten. Gluten is a protein which is found in stronger flour, and develops into long structures as the flour is worked. It will harden when cooked, trapping any air bubbles. This can be seen when making bread. Self-raising flour creates a softer finished product, and is used in baking cakes.

- You would then select three of the flours that you wish to use. Make sure you that you have detailed explanations of what these three flours are, what they contain and what their properties are. Use several different sources for your information – for example reference books or websites. Make a note in your work of where you got the information from (this is called **referencing** your work).
- You would give a list of examples of when these flours are used in recipes and discuss each recipe, using scientific terminology when you can to explain what the role of the flour is in each recipe.

  An example of this could be:

> Strong bread flour is used in this selected bread recipe because it has high gluten and protein content. It is made from hard (or winter) wheat, which is sown in the autumn and after germination it lies dormant over the winter and resumes growth in the spring.

> *Gluten is the protein that is stretched during the kneading process while making bread. It provides a structure to the bread, and traps the carbon dioxide bubbles that are produced when the yeast is growing during the proving and rising of the dough.*
>
> *When the bread is cooked, the protein that is gluten is denatured by the heat from the oven, and it bakes into a solid structure that produces the bread.*

- You would then need to narrow down your selection to three recipes and explain in detail how the property of each type of flour is being investigated in your choice of recipe.
- Make a plan of action. State your **hypothesis**. This means you have to write about what you think will happen when you make and cook your chosen dishes. Make this as detailed as you can for the top marks. Use scientific words if you can. An example of this for the sample task could be:

> *I think that the plain white flour will produce the best result in the shortcrust pastry because it is a soft flour that does not contain much gluten. Shortcrust pastry needs to have a crisp bite to it, so the soft, plain flour will allow the fat to coat the flour during the rubbing-in process, preventing long strands of gluten from forming. I do not want the pastry to rise during cooking, so as plain flour does not contain raising agents, this will not happen.*

- You will then plan to make each of the three recipes, using each of the three flours. This will mean that you are making nine products in total. *(When making your recipes, you do not need to make full, family-sized products – for example, if making shortcrust pastry, you could follow a recipe simply to produce a small sample of the pastry, e.g. 100g flour, 50g fat.)*

For the highest marks in this section you will have to show that:
- You fully understand the task by having a detailed interpretation.
- You have detailed research from a number of different sources.
- You have produced a comprehensive plan, considering a number of different recipes and hypotheses for each choice given.
- You have used all scientific terminology appropriately.

Section B: Carrying out your plan of action (15 marks)

This is the section where you will do the practical work, which must include trialling of recipes, developing and testing, andlooking at the functional and chemical properties of the ingredients that you have selected to use.

You will need to describe the procedures that you carried out in detail, explaining what you did.

You must carry out taste testing using preference tests with a panel of tasters, using sensory testing, charts, and customer feedback via peer assessment.

Take photographs of all your products to provide evidence for the examiner to see, including during the making process.

Figure 13.2 Risen bread product

- For the example task, begin by making the first of your recipes with each of the three flours. You may have chosen to make a bread product which needs to rise, and therefore requires a strong flour which contains gluten to make the structure of the bread. You would make the bread product using strong white bread flour, plain flour and self-raising flour. Your hypothesis for this recipe would be similar to the example given above. The plain flour and self-raising flour do not contain as much gluten, so the outcome from these flours will not rise as much, or provide a strong enough structure to support the air bubbles when the bread is cooked.

- Make the three recipes, taking photographs during the making and of the final outcomes.

- Set up preference testing with a tasting panel of your peers. (Details of different types of preference testing are provided in Chapter 10 Factors affecting food choice – see pages 184–203. You may find it useful to refer back to this chapter to help you.)

- Note down all the preferences, and enter the results on a chart of your choice. Use an IT program such as Excel if possible to produce the charts.

- Now continue with the other two recipes in the same way.

For the highest marks in this section you will have to show that:
- You have trialled and modified the ingredients while following your plan of action during the preparation and cooking processes. This may mean that you have to change the type of fat you are using or the brand of flour.

- You have an excellent detailed understanding of the working characteristics and chemical and functional properties of the ingredients you are using, and your final choices have been based on your understanding.

- You have recorded in detail any changes that you have made during the preparation and cooking process. An example of this may be that you are using a reduced-fat spread that contains a high percentage of water, so the outcome was not successful.

- You have carried out and recorded detailed testing of your products. You need to set up your preference testing in a way that allows the tests to be fairly carried out. You need to explain what you have done to ensure fairness of testing. It may be that you have only allowed one person at a time into the room to test the products, and that testers are not allowed to discuss their preferences with one another.
- You have photographed all stages of your preparation and cooking, and have annotated each photograph with detailed explanations. You have explained carefully what each photograph shows.
- You have presented your results clearly and logically using a range of different formats. This could include different types of charts, such as pie charts, histographs and spider diagrams.

Section C: Analyse your data and results (10 marks)

This is the section where you will report on the outcomes and results of your preference testing. You will justify what you did and why your recipe worked or did not work, explaining the reasons why. You will say whether your hypothesis was correct or not.

The analysis of your results must be clear and be backed up by your research. You must refer back to your hypothesis or educated guess. State why the recipe was or was not successful, using scientific words when necessary.

For the highest marks in this section you will need to show that:
- You have analysed the task in detail, and you have made conclusions that are based on scientific knowledge and understanding of the ingredients that you chose to use in your task, and the preparation and cooking methods that you used.
- You have an excellent understanding of the task – you have justified what you have found out with clear reasons based on the methods you used to gather the results of your testing.
- You have looked at the hypothesis you made at the beginning of the task, and you have reviewed it using the data and information from your research and investigation, showing that you understand the task.
- Your writing is clear and easy to understand, and you have used scientific words and terminology in your work.

An example of how you would report on the outcomes and results of your preference testing of making the bread product would be:

> *The photographs of my bread products show that the best outcome was from using the strong bread flour. When the bread was sliced open I can see that the texture of the bread is even and there are no large holes in the dough.*
>
> *In my hypothesis, I stated that the strong bread flour contained the most gluten and therefore the finished product using this flour would be likely*

to be the most successful as during the making of the bread product, the gluten would be stretched during the kneading process. This means that it provides a structure to the dough, which will allow the air bubbles produced while the yeast is growing to be trapped, allowing the dough to rise. If the bread has been sufficiently kneaded, the dough will rise evenly and once the bread is cooked, the gluten denatures and hardens, trapping the air in the cooked bread product. The finished bread will be light, evenly cooked and when it is cut will not have any large air holes in the dough.

The bread products made with the plain flour have not risen as well as the one made with strong flour. The product also has an uneven rise, and when it was sliced open the dough was compressed and unattractive. This is because the flour does not contain much gluten, and was unable to form long strands to support the dough during the proving stage of making the dough when the yeast is growing and producing carbon dioxide. This has resulted in an uneven rise and a stodgy final product.

The bread product made with the self-raising flour has also produced an unsuccessful final outcome.

The final product has not risen evenly, and is much smaller than the one made with strong bread flour. Again this is because the flour does not contain much gluten, and the dough is unable to support the bubbles of carbon dioxide that is being produced by the yeast during the proving stage of making. The self-raising flour also contains its own raising agent in the form of bicarbonate of soda and baking powder. These products will counteract the action of the yeast, as they will work to make the product rise. This shows in the uneven rise of the product using this flour.

During the taste testing I presented all three products in the same way, serving the bread with butter and jam. This was to ensure that each person tasting the products had exactly the same experience of tasting, providing a fair test.

The results of my preference testing show that the product using the strong bread flour was the favourite product chosen by the people who taste tested the three products.

You would continue to discuss and analyse all the products that you have made in a similar way, going back to your research and referring to the types of flour, what they are made from, what they contain that makes your products successful, or not.

CHAPTER 14
Assessment 2: Food preparation assessment

Create dishes to form part of a menu

The second assessment will be a food preparation assessment.

You will need to include evidence to show that you can:

- **interpret and investigate** your chosen option
- **showcase** your technical skills
- **evaluate** your completed practical assessment and food outcomes.

This is the assessment where you are showcasing your technical skills.

- You will need to produce a completed project of a maximum of **15 pages, or 30 sides, of A4** paper (or A3 equivalent), to include all photographs, charts and graphs, using **font sized 11 or 12.**
- You will have a maximum of **12 hours** for this assessment: **8 to 9 hours** for the research, planning, trialling and evaluation, and **3 hours** in a single cooking session to produce a minimum of three dishes with appropriate accompaniments to fit the task given.
- You will be expected to work hygienically and safely during this assessment.

This assessment is 35 per cent of your total GCSE marks.

You will get a choice of **two** different assessment tasks.

In this assessment you will have to:

- Carry out research into your chosen task, trial and test selected recipes, produce a final menu and create a plan of action for your cooking. There is a maximum of **15 marks** for this section.
- Prepare, cook and serve a minimum of three dishes with appropriate accompaniments in a single cooking session of three hours. There is a maximum of **45 marks** for this section.
- Evaluate the selection, preparation, cooking and presentation of your dishes. There is a maximum of **10 marks** for this section.

An example of an assessment could be as follows:

> **You are trying to encourage teenagers to eat more fruit and vegetables. Research, prepare and cook a selection of dishes that show the use of fruit and vegetables, and produce three dishes that will appeal to teenagers.**

Section A: Research, trialling and testing, selecting a menu and making a dovetailed time plan (15 marks)

- For the example task, you would begin by looking at why teenagers need to be encouraged to eat more fruit and vegetables (chapter 4 provides information on planning balanced diets for different groups of people). Explain what nutrients are contained in fruit and vegetables and what these nutrients do in our bodies.

- You would then carry out a survey of teenagers to find out how much fruit and vegetables they eat. Present your results in graphs or pie charts and annotate and explain what the results of your survey shows.

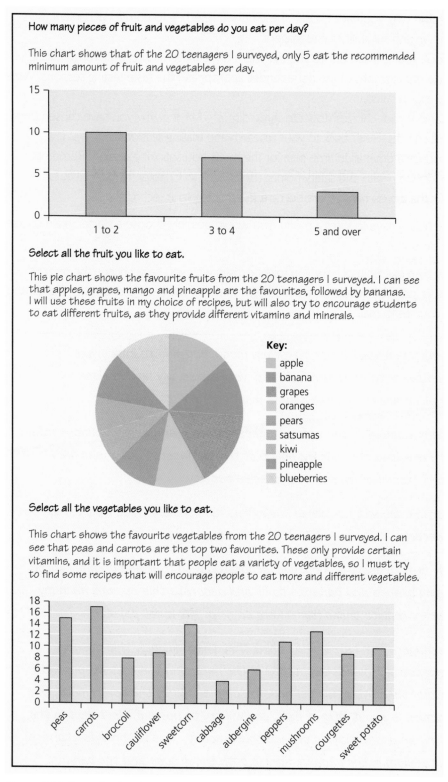

How many pieces of fruit and vegetables do you eat per day?

This chart shows that of the 20 teenagers I surveyed, only 5 eat the recommended minimum amount of fruit and vegetables per day.

Select all the fruit you like to eat.

This pie chart shows the favourite fruits from the 20 teenagers I surveyed. I can see that apples, grapes, mango and pineapple are the favourites, followed by bananas. I will use these fruits in my choice of recipes, but will also try to encourage students to eat different fruits, as they provide different vitamins and minerals.

Key:
- apple
- banana
- grapes
- oranges
- pears
- satsumas
- kiwi
- pineapple
- blueberries

Select all the vegetables you like to eat.

This chart shows the favourite vegetables from the 20 teenagers I surveyed. I can see that peas and carrots are the top two favourites. These only provide certain vitamins, and it is important that people eat a variety of vegetables, so I must try to find some recipes that will encourage people to eat more and different vegetables.

Figure 14.1 Example of student-presented survey results

- Produce some ideas for dishes that you could cook. Remember to show a selection of skills when choosing your recipes. Refer back to the list of 20 skills that you need to show as detailed in Table 11.3. Include as many of these as possible, using your saved portfolio of recipes that you know you can produce successfully.
- Explain why you have chosen these dishes, referring to nutritional content, skills shown and suitability for teenagers.
- Cook your chosen dishes and trial them on teenagers using preference testing. Take photographs of the dishes and, if appropriate, the taste testing sessions. Write up your results.
- Narrow your choices down to three dishes, justifying why you have chosen these final choices. Refer back to your research and trialling sessions.
- Produce a dovetailed time plan for the three-hour cooking session. Remember to include health and safety points. (Refer back to Chapter 11, page 209, for information on how to write a dovetailed time plan.)

As the dishes have to form a menu, you would possibly choose a starter, main course and a dessert.

Below is an example of how you might explain why you have chosen the dessert dishes to trial. You would have already written why you have chosen your starter and main courses to trial.

I had to choose dishes to form a menu, but also to encourage teenagers to eat more fruit and vegetables. My final choices for trialling are for a three course meal.

Many teenagers tend to eat unhealthy food, and as you can see from the results of my survey, only 5 of the teenagers I asked eat five or more pieces of fruit and vegetables per day.

For the dessert section of the menu I have chosen three fruit based desserts.

My first dessert is a sponge, decorated with fruit and cream. I will be using grapes and bananas to fill and decorate this sponge, as in my survey results these fruits were some of the most popular.

Teenagers love cakes, and I think that they will enjoy this dessert because it provides sweetness in the cake, and the fruit will provide two of their five a day portions recommended by the government. One portion is about 80g of fruit, so I will try to use sufficient fruit in the filling an decoration to provide this amount. This will provide vitamins, potassium in the banana, fibre in both the grapes and the banana.

I will also use cream as a filling with the fruit. This will provide a nice soft filling for the sponge, and as it contains fat it will give a creamy mouth feel to the product. I will use half fat cream, as the sponge cake contains fat and sugar which gives it a high calorific value.

My second dessert is a mille-feuille filled with blueberries and yoghurt. This is made with puff pastry layers, which are sandwiched together with the fruit and yoghurt. Blueberries were one of the most popular fruits in my survey. They provide a lot of vitamin C, which is important for healthy skin and gums. It also helps with the uptake of iron form the foods we eat. Teenagers, particularly girls, can lack iron. The yoghurt provides some calcium which is important for healthy bones and teeth. Teenagers are still growing, so they need calcium.

I think this will be a popular dessert, and I will be making the puff pastry which shows a high level skill.

My third choice of dessert for trialling is a filo pastry basket filled with a selection of fruit, drizzled with chocolate. I will fill the basket with sliced apples and grapes, as these were popular fruits in my survey.

I will be making the filo pastry. This dessert contains less calories than the sponge and the mille feuille as the pastry is rolled very thin. It may be a better choice to go with my starter and main dish choices, as it is not such a filling dessert.

By drizzling it with a thin layer of chocolate, I hope to appeal to the teenagers, who usually love chocolate.

My first dessert is a creamed sponge cake with whipped cream, grapes and bananas.

I am not very pleased with the texture and appearance of the sponge, but the people who tasted my sponge said it tasted delicious.

The results of the taste testing are set in a graph below.

I asked the people tasting my sponge to give marks out of ten for appearance, taste and texture.

Appearance	Taste	Texture
6	8	7
5	8	8
6	9	8
6	7	7
5	8	7
6	8	8
6	9	9
5	7	8
6	8	8

As you can see, the appearance was not given very high marks. I think I could have been more careful cutting and decorating the sponge. When I make my final sponge, I will make some changes.

I will change the sponge to a whisked sponge, as I think that this gives an easier texture sponge to cut and it will look better when sliced.

I will also use blueberries as well to add more colour.

My second dessert is a mille feuille, made with home-made puff pastry, and filled with thick Greek yoghurt and blueberries.

I am very pleased with the rise and layers that I got on the puff pastry.

I think that this looks very attractive.

The results of the taste testing are below.

Appearance	Taste	Texture
8	8	8
9	8	9
9	7	9
9	9	9
8	8	8
8	9	8
9	8	9
10	9	10
9	9	9

As you can see, this dessert scored high marks in all three categories. I am really pleased that this was so successful.

However the puff pastry took a long time to make, as it needs to be rested between the rolling and folding.

I am not going to choose this dessert for my final choice because of the time factor, as I only have three hours to complete and serve all my dishes.

My third dessert is a filo basket, filled with grapes, bananas and blueberries, and drizzled with chocolate.

This looks very good, and I am pleased with the final outcome.

The marks for the taste testing are shown below.

Appearance	Taste	Texture
10	8	8
9	8	8
9	9	9
10	8	8
9	8	8
9	9	8
9	8	9
9	8	8
9	8	8

As you can see, this was also a success with the people tasting the dessert. The pastry was very crisp and the chocolate added sweetness to the dish.

The filo pastry took a very long time to make, and rolling it out so thinly was difficult.

I will not be choosing this dessert because of these difficulties.

Narrowing down your choices

When you are selecting your final three dishes to cook, an example of how you would write your **justification** for your choice may look something like the example below. This is a justification for the **dessert** selection from the three choices you trialled. You would have already written justifications for your selection of **starters** and **main courses** to trial.

I tried out the fruit sponge, mille feuille and filo baskets with my chosen testers. As you can see from the results of my taste tests, the fruit sponge was the most popular.

I think this is because most teenagers are used to eating cakes, rather than pastry products. Some of the people that tried my desserts had not eaten filo pastry before, as it is not that common, and were unsure as to whether they would choose it as a dish to eat again. They did enjoy the filling.

I tried out the fruit sponge, mille feuille and filo baskets with my chosen testers. As you can see from the results of my taste tests, the fruit sponge was the most popular.

I think this is because most teenagers are used to eating cakes, rather than pastry products. Some of the people that tried my desserts had not eaten filo pastry before, as it is not that common, and were unsure as to whether they would choose it as a dish to eat again. They did enjoy the filling.

I am going to change the recipe of the sponge from a creamed sponge to a whisked sponge. This will not only reduce the fat content of the sponge, making it a more healthy option, but will also reduce the cooking and cooling time of the sponge.

The filo pastry and the puff pastry took a long time to make, and when I looked at the other dishes that I have to cook, I decided that as I only have three hours to make and serve all three dishes, I would also select the fruit sponge, as it does not take so much time to cook and decorate. The cake only takes ten minutes to make, and while it is cooking in the oven I can get on and make my other two dishes.

I will be adding blueberries to the fruit for my sponge for the final dish, as all the teenagers who tried my desserts stated that they thought the filling for the filo baskets tasted delicious. I will drizzle some chocolate over the top of the sponge, as the comments on the filo baskets included reference to the lovely combination of the blueberries and chocolate.

This means that my final dessert choice will be a sponge cake, filled with sliced bananas and blueberries in half fat whipped cream. The top will be decorated with cream, sliced grapes and blueberries and then it will be drizzled with melted chocolate.

While the sponge is quite high in fat and sugar, when you add the fruit, you can reduce the size of the portion of cake because I am adding three different fruits. This means that the size of the cake portion does not have to be so big, as the fruit will make up a large part of the dessert.

These are pictures of my final dessert.

For the highest marks in this section you need to do the following:

- Demonstrate an excellent understanding of the task, use different research techniques such as primary research (direct questioning of teenagers, for example), and secondary research, and use at least three different sources of information for your research.
- Use trialling and testing of your initial choices to make a final selection of dishes.
- Select dishes that allow you to fully showcase a range of the more difficult skills from the list of 20 in Table 11.3.
- Fully justify your choices, using your research and trialling results.
- Produce a clear, dovetailed sequence or plan of work that includes referencing of safety and quality control points. Your plan of work should be able to be followed by someone else without you being present to produce excellent quality outcomes.
- Use the correct scientific and cooking terminology to describe what you are doing.

This section should be 15 to 17 sides of A4 paper (or A3 equivalent).

Section B: Showcasing your technical skills (45 marks)

This is where you must show the examiner that you can carry out your chosen recipes successfully and competently, finishing up with excellent presentation skills and well-cooked dishes.

For the highest marks in this section you need to:

- Follow your time plan accurately and work independently and competently during the preparation, cooking and presentation of your dishes.
- Use the three hours that you have been given fully, using all the correct equipment for each dish, and preparing each dish with confidence.
- Show a wide variety of skills for each dish, including complex skills such as filleting fish or precise cutting of vegetables, to an excellent standard.
- Manage the time well and produce all three dishes successfully in the three hours.

Figure 14.2 A student preparing ingredients in a kitchen

- Show that you are checking the sensory properties, such as seasoning, by tasting and adjusting throughout the cooking process.
- Show that you are adjusting temperatures during the storing/cooking/testing so you can serve the dishes at the correct temperature.
- Present your final dishes to an extremely high standard, accurately portioning your dish, and showing how it would form part of the meal.
- Decorate or garnish your dishes to show that you can pay attention to detail to improve the aesthetic qualities of each dish.

You must take photographs of your final dishes to put into your assessment.

The photographs should take two to three pages of A4 paper (or A3 equivalent), and be annotated fully.

Section C: Evaluating the Assessment (10 marks)

In this section you will need to show that you are able to evaluate the technical skills that you have selected and demonstrated during your cooking, and that you can use sensory descriptors to consider the taste, texture, aroma and appearance of your dishes. You also need to evaluate the presentation and food styling of the completed dishes.

For the highest marks in this section you will need to:
- Fully evaluate the technical skills that you have used, explaining why you carried these out for that particular dish. An example could be:

> *I filleted the fish before I cooked it as the dish I chose needed to be made into four equal-sized portions. I wanted to roll the fillets around the mushroom stuffing as during the trialling of this dish I found the best presentation could be achieved by preparing the fish in this way.*

- Evaluate your dishes to a high standard, using accurate sensory descriptors for the taste, texture, aroma and appearance.
- Detail any changes or modifications you made to the original recipes and explain how this has improved the success of each dish.
- Show an understanding of the nutritional content of your dishes, and relate that to current government guidelines for healthy eating.
- Explain how the sustainability of your chosen dishes, including reference to food miles, influenced your choice.
- Use the correct technical terminology with ease, look at other students' work and state how your finished dishes compared to other students' work.
- Make detailed suggestions for realistic improvements to your own work, after you have compared it with other students' work.

This section should be eight to ten sides of A4 paper (or A3 equivalent).

Evaluating the technical skills.

This example shows how you might evaluate the skills you used to produce the fruit sponge for your dessert choice, you will already have written an evaluation of the skills for your **starter** and **main course**.

When making my sponge, I showed the skills of weighing and measuring accurately. This is important, because if incorrect amounts of ingredients are used, the outcome will not be successful.

I also showed the skill of incorporating air during the whisking of the sponge mixture. When I whisked the eggs and sugar together, the action of whisking them together results in air being trapped in the mixture. The protein in the eggs is also denatured during the whisking, and is stretched. This means it will set around the air bubbles when the sponge is cooked, helping to create the structure of the sponge. I sieved the flour before I added it, to remove any lumps, and to help add air to the flour.

When I added the flour, I made sure I folded it in carefully, using a figure of eight movement with a metal spoon, to avoid knocking any air out of the mixture.

I greased my cake tins, and lined them with greaseproof paper so the mixture did not stick in the cake tins, and the final sponge was perfectly cooked and looked professional.

I made sure I spread the mixture out evenly between the two cake tins and made sure the mixture was level in each tin so that each cake was a similar size and cooked in the same time. This also gave an even rise to the cakes.

I preheated the oven so it as the correct temperature before the cakes were put in.

I left the cake to cook for fifteen minutes before I checked and opened the oven door to see whether it was cooked. This was important, as if the door was opened before the cake was cooked, it could cause the cake to collapse.

When decorating my cake, I sliced the banana thinly and placed it in lemon juice before I used it to fill the cake, as this will help prevent the banana discolouring due to enzymic browning.

> *I whipped the cream and then used some of it to fill the cake and the rest to cover the top of the cake. The grapes were cut in half, and the blueberries were washed, dried and used to decorate the top of the cake.*
>
> *I melted the chocolate in a bowl over hot water, to prevent it burning, and then drizzled it over the cake.*

Evaluating your dishes to a high standard using sensory descriptors.

When you evaluate your dish, remember that the examiner who is marking your work will have to rely on your photographs and your sensory descriptors to decide whether your dish is of a high standard and deserves high marks.

You will already have written an evaluation of your **starter** and **main course**.

This example shows what a written evaluation of the fruit sponge could be. Use table 11.4 with the lists of sensory words in Chapter 11. You can also use tables, charts and star diagrams to show the marks that your tasters gave. You can then annotate these to explain what they show, to inform the person reading your work about the sensory analysis of your dish. You can find more information about star profiles and how to present taste test results on pages 188 and 189.

> *My fruit sponge looks appetising, attractive and colourful. The sponge is well risen, and the two different fruits on the top add colour, green from the grapes and the dark purple/blue of the blueberries. The drizzled chocolate makes a nice attractive contrast to the pale cream.*
>
> *When my sponge is cut it has a springy texture when the knife goes in. This is because of the well risen airy sponge, which slightly pushes against the knife as it cuts through. It has a close spongy texture, that has no obvious large holes in, so it was well mixed, with the eggs and sugar being beaten until they were pale and thick enough to leave a trail in the mixture, and with all the flour being evenly incorporated.*
>
> *It has a lovely golden yellow colour, and the top was golden brown due to the dextrinisation of the sugar during cooking, and the egg contributing to the golden brown colour.*
>
> *When you bite into the sponge you get a contrast between the soft sponge, the slight bite of the sliced banana and the creamy filling, which melts in your mouth. The blueberries and grapes on the top of the sponge provide a firm bite to contrast against the soft sponge and creamy filling.*

> My sponge tastes sweet, with a lovely creamy mouth feel from the whipped cream. The bananas were coated with lemon juice to prevent enzymic browning, and the slight tang of the lemon juice gives a sharp flavour to the fruit, which is quite refreshing. The blueberries and grapes release a wonderful fruity flavour which bursts into your mouth as you bite into the sponge topping. This is contrasted by the creamy flavour and mouth feel of the whipped cream which is under the fruit. The chocolate gives sweetness and a slightly fatty flavour, which coats the inside of the mouth.

Show an understanding of the nutritional content of your dish:

You will need to explain how your choice of ingredients is relevant to the task that you have been given, and state which nutrients are found in your choice of dish, and what they do in the body.

The example below shows how you could write about the **dessert** course. You will already have written about your **starter** and **main course**.

In this case it is the fruit that you are using, but you can also comment on the nutritional content of the other ingredients in the fruit sponge.

An example of part of this section could be:

> My task was to try to encourage teenagers to eat more fruit and vegetables by designing and making a menu that is suitable for a teenager.
>
> My fruit sponge will contribute to the overall daily nutritional needs of a teenager, while being tasty and attractive enough to encourage a teenager to eat it.
>
> The sponge contains sugar, eggs and flour. I changed the sponge from a creamed sponge to a whisked sponge in order to reduce the fat content of the sponge. Fat is important as an energy source, for providing an insulating layer in our bodies and for protecting our body organs. It also can contain vitamin A and D which are fat soluble vitamins, if they have been added to the fat, for example in margarine. However too much fat can cause problems in the body. It can cause someone to be overweigh if too much is eaten and it is then stored in the body. Saturated fats can also cause problems by raising the cholesterol levels in the blood, contributing to heart disease. The recommended amount of fat for an average man is 95g a day, of which not more than 30g should be saturated fat.

Many teenagers seem to prefer fatty, sugary foods, so I am reducing the fat to try to help them stay within the recommended limits.

The eggs in the sponge are a source of protein, which is needed for growth, repair and to make hormones and enzymes in the body.

The flour is carbohydrate which will provide a little amount of energy.

The sugar is also carbohydrate. Too much sugar will be bad for the body as it can cause obesity and dental caries.

The fruit I am using is banana, grapes and blueberries.

The bananas are a source of potassium which is needed to help build proteins, to break down carbohydrates during digestion, to build muscle and to control the electrical activity of the heart.

The blueberries are an excellent source of anti oxidants, which help mop up free radicals in the blood which can cause some cancers. They also contain Vitamin C which is needed for healthy skin and gums and to help with the absorption of iron from other foods.

The grapes contain small amounts of Vitamin B1 (Thiamine), which is needed for releasing energy from carbohydrates during digestion, helping the nerves in the body to work properly and helping with growth in the body. The grapes also contain Vitamin B2 (Riboflavin), which is needed for helping release energy from carbohydrates during digestion and keeping your skin healthy.

Grapes contain Vitamin B6, which is used in digestion to break down carbohydrates, and they also contain Vitamin K, which helps blood to clot.

All the fruit used here contain fibre which is needed to speed up the digestive process and prevent constipation by adding bulk to the food, making it easier to pass through the digestive system.

Explain how the sustainability of your chosen dish, including reference to food miles, influenced your choice.

You will need to think about how far the food that you have chosen has travelled. If you have use ingredients from local sources, state this. If you concentrated on using British products, include this information.

An example of the explanation for the fruit sponge is show below. You would have already explained about your **starter** and **main course**.

WJEC EDUQAS GCSE Food Preparation and Nutrition

I am very aware of the impact of food miles on the environment, and try to use local food whenever possible, to reduce the carbon footprint of the ingredients that I buy.

There is a local farmers market in my town, so I shopped for the eggs here, which came from a local farm. This means they will not have travelled so far. This provides income for the local farmer, which means that he or she will be able to spend money locally, and hopefully employ local people to work on the farm as the money spent on their eggs will stay in the local area.

The sugar is a British product, and states clearly on the pack that is it grown and produced in Britain. This provides employment for British people, and generates money through taxes for the economy.

The flour is also a British product, grown and milled in Britain, so the same reasons apply.

The blueberries were bought at a local supermarket. There was a choice of blueberries from Peru, and blueberries from Spain. I chose the blueberries from Spain as they had not travelled so far.

The bananas were also from the local supermarket. They come from Ecuador, but are Fair Trade. This means that although they have travelled a long way, the farmer was given a fair price for his crop. They are slightly more expensive than non Fair Trade bananas.

The grapes come from Spain, which again is creating pollution when the food is transported, but it is not possible to get grapes that are grown in the U.K.

Use the correct technical terminology with ease, look at other students' work and state how your finished dishes compared to other students' work.

When you are working with other students in your non exam assessment, it would be a good idea to look at other students work after you have completed your work. You can see how their outcomes compare to your outcomes.

Make sure you are using all the correct words for the comments you have made on nutrition content, the functions of ingredients in your recipes and that you use sensory descriptors when evaluating your practical work.

The example below shows how you could compare your **dessert** to another student's dessert. You would already have compared your **starter** and **main course** to another student's starter and main course.

> I am very happy with the outcome of my dessert. I have looked at the desserts produced by other students and have noted that some of the students have not used as many skills in the products they have produced.
>
> An example of this is that one person has made a strawberry mousse, which only required them to put the ingredients into a blender, then decorate with fresh strawberries.

Think about other comparisons you could make, for example the appearance of the outcomes, or the nutritional content.

Make detailed suggestions for realistic improvements to your own work, after you have compared it with other students' work.

You need to suggest several realistic improvements to either improve the appearance, taste, texture or nutritional content of your completed dish.

An example of suggestions for your **dessert** is shown below. You will already have suggested improvements for your **starter** and **main course**.

> My dessert looks very attractive. I could improve the appearance by piping rosettes of cream on top around the edge, and making it look more professional in this way.
>
> Instead of drizzling the chocolate over the sponge I could make little chocolate curls, or shapes to place on top of the cake.
>
> I could change the fruit that I am using or add extra fruit such as oranges, to give more colour and extra Vitamin C.
>
> I could replace the cream with Crème Fraiche, which would further reduce the fat content of the dish.
>
> I have seen that one of the other students has made a whisked sponge, but instead of a round sponge they made a Swiss roll, which was loaded with fruit. This looked very attractive, and could be a different option for me to choose.

SECTION 8

The written examination: Principles of food preparation and nutrition

This section includes the following chapter:

Chapter 15 The written examination: Principles of food preparation and nutrition

The written examination is **50 per cent** of your final GCSE qualification. This examination is available in both paper-based and online formats.

There will be **100 marks** available.

You will be tested on all the topics that you have covered over the course. These are:
1 **Food commodities**
2 **Principles of nutrition**
3 **Diet and good health**
4 **The science of food**
5 **Where food comes from**
6 **Cooking and food preparation**

All of these topics are fully covered in Sections 1–6 of this book.

The exam paper will be divided into two sections:
- **Section A** will be worth **15 marks** and will ask you to answer questions related to a visual stimulus, which will be included on the exam paper.
- **Section B** will be worth **85 marks** and will include different question types.

Assessment objectives

Your responses in the exam will be assessed against the assessment objectives (AOs) shown in this table.

	Assessment objective (AO)	Weighting for written exam
AO1	Demonstrate knowledge and understanding of nutrition, food cooking and preparation.	20%
AO2	Apply knowledge and understanding of nutrition, food, cooking and preparation.	20%
AO3*	Plan, prepare, cook and present dishes, combining appropriate techniques.	0%
AO4	Analyse and evaluate different aspects of nutrition, cooking and preparation including food made by themselves and others.	10%
*You will not be assessed against AO3 in the written exam, only AOs 1, 2 and 4.		

Some questions near the end of the paper may be assessed against more than one of these assessment objectives, and may be marked in 'bands'. For these questions, you may have to give more answers than the number of marks shown in order to get maximum marks. For example, you may need to give six or eight points to earn four marks. Some advice on answering these types of questions is included on the following pages.

Advice for answering exam questions

- **Read** all the instructions on the front of the paper. Make sure you understand how long you have to answer the questions. You will need to answer **all** questions.
- **Check** you are using the correct colour pen. You need to use a **blue** or **black** pen. Do not use pencil or gel pen.
- **Read** the question through **twice** before you start to answer. Make sure you understand what you are being asked to do.
- **Underline or highlight** key words to help you focus on what you need to write.
- **Check how many marks** are given for the question. That should tell you how many points or pieces of information you need to include in your answer (although remember that some questions are marked in bands and may need you to provide more points than the number of marks shown).
- **Understand** what the **key words** in a question mean. Different words require a different type of answer.

This table shows you what different words mean in a question.

Key word	How to answer the question
Identify/suggest/give a reason for	Make a list, write a short answer, select words from a diagram or table to complete gaps in a sentence.
Describe	Make a detailed explanation as to how and why something happens.
Explain	Clarify a subject or point by writing down the meaning of it and then showing you understand it by giving reasons.
Analyse	Break an issue down into its separate parts and look at each part in depth, using evidence and explanations to show your understanding.
Evaluate	Make a judgement about how successful or unsuccessful something is and say why it is important. Include evidence for your answer, and come to a final conclusion.
Discuss	Write about all evidence for and against a topic, or point out the advantages and disadvantages of a topic. Use evidence to arrive at a conclusion.

Table 15.1 Key words in exam questions

Types of questions

The exam paper will be a mixture of questions, using the words from the table above. This means that each question will need to be answered in a different way.

Data response questions

These questions will give you a table of data or a chart, pie chart or graph.

You will be asked to extract information from the data and answer questions on it. You may also be asked further questions that show your understanding of the topic. Each part of the question will have a different number of marks.

An example of this type of question is as follows:

1   A survey was taken of three different age groups to find out how many pieces of fruit and vegetables the students ate per day.

30 students in each age group were given a food diary to complete in a week. The results were added up and the average number of fruit and vegetables per day were calculated.

The table shows the average daily consumption for each age group.

Look at the table then answer the following questions.

Pieces of fruit and vegetables	5 to 7 years old	8 to 10 years old	11 to 15 years old
1 piece	4	9	10
2 pieces	6	5	6
3 pieces	10	6	4
4 pieces	6	8	5
5 pieces	3	2	1
more than 5 pieces	1	0	0

Table 15.2 Average pieces of fruit and vegetables eaten by students of varying ages per day

(a) How many 5- to 7- year old students ate less than five pieces of fruit and vegetables per day on average? [1 mark]

(b) How many 11- to 15- year old students ate more than three pieces of fruit and vegetables per day? [1 mark]

(c) Give **two** reasons why you think that no student over 8 years old ate more than five pieces of fruit and vegetables per day. [2 marks]

(d) Suggest **three** ways in which you could encourage 8- to 10- year old students to eat more fruit and vegetables. [3 marks]

(e) Explain why it is important to eat a minimum of five portions of fruit and vegetables per day as recommended by the government. [6 marks]

To answer this question:

- **Refer** back to the data, explaining how you got your answer, so the examiner can see that you understand.
- **Use the data** plus your own knowledge to answer the questions that require a fuller response.

Some sample answers to this question are shown below, along with some commentary on how you might be awarded marks.

(a) Answer: 26

Commentary: There is just one mark available for this question so you only need to give a single answer by adding up those 5 to 7 year olds who ate 1, 2, 3 and 4 pieces of fruit and vegetables (i.e. less than 5).

(a) Answer: 26

Commentary: Again, there is just one mark available for this question so you only need to give a single answer by adding up those 11 to 15 year olds who ate 4, 5, or more than 5 pieces of fruit and vegetables (i.e. more than 5).

(c) Answer: *Their parents may not buy much fruit so it is not in the house for them to eat. Their parents may be on a low income so cannot afford much fruit.*

One mark is awarded for this sentence

One mark is awarded for this sentence

Commentary: In this question you are being asked to apply your knowledge and understanding of the factors that affect food choice. There are two marks available for this question and you are asked to give two reasons, so you are expected to give two points or facts. The student here gets one mark for each point they give. Other answers they could have given include:

- They do not like fruit very much so it is not in the house for them to eat.
- Their friends do not eat fruit so they do not eat it.

One mark is given for each suitable answer (up to the maximum two marks available). A mark will be given for any suitable answer, even if it is not one of the answers given above.

(d) Answer: *You could add fruit to desserts that they like (for example, fresh strawberries to decorate a chocolate mousse). You could add dried fruit to cakes and biscuits. You could also disguise vegetables in dinners (for example, making a homemade pasta sauce with blended vegetables in it).*

One mark is awarded for this sentence

One mark is awarded for this sentence

One mark is awarded for this sentence

Commentary: In this question you are being asked to apply your knowledge and understanding of fruit and vegetables and how to consider nutritional needs and food choices when developing recipes and meals. There are three marks available for this question and you are asked to **suggest three ways**; you are therefore expected to make three points. The student gets one mark for each answer they provide. Other ways of they could have suggested include:

- Make fresh fruit smoothies
- Add vegetables to pizzas or blended with burgers
- Dip fruit in chocolate or yoghurt to make it sweeter and more attractive
- Have cut up vegetables in the fridge for a snack with a dip

A mark will be given for any suitable answer, even if it is not one of the answers given above.

One mark is awarded for this point.

One mark is awarded for the explanation of the function of vitamin C.

One mark is awarded for stating that fruit and vegetables are a good source of fibre; another mark is awarded for explaining the function of fibre.

(e) Answer: *Fruit and vegetables provide vitamins and minerals such as Vitamin C, which is responsible for healthy skin and gums, and helps non-haem iron to be absorbed from plant sources of iron. Fruit and vegetables are a good source of fibre which helps the digestive system to work and prevents constipation. Dark green vegetables are a good source of iron, which is needed for making red blood cells which carry oxygen around the body.*

One mark is given for stating that green, leafy vegetables are a good source of iron; another mark is awarded for explaining the function of iron.

Commentary: In this question you are being asked to apply your knowledge and understanding of recommended nutritional guidelines for a healthy diet when planning balanced meals, and your knowledge and understanding of the nutritional value of fruit and vegetables in the diet. The question asks you to **explain** – you therefore need to give reasons for each point you make about why eating a minimum of five portions of fruit and vegetables per day is important. There are six marks available for this question – you will gain one mark for each correct point you make, and an additional mark for giving a reason for each point. You should therefore provide three points with a reason for each to gain the maximum six marks. The answer above would be awarded full marks. Other answers could have included:

- Calcium is found in dark green leafy vegetables. Calcium is needed for strong bones and teeth.
- Magnesium is found in dark green leafy vegetables. Magnesium helps to support the immune system.
- Fruit and vegetables contain antioxidants that fight against disease and can help prevent cancers by mopping up free radicals.

You would gain marks for any other vitamin mentioned with its function.

Now have a go at answering the second question below.

2 Look at the food label for a pack of **two** English muffins.

   (a) What percentage of the muffin in total is carbohydrate? [1 mark]
   (b) How many grams of fat are included in this product? [1 mark]
   (c) What is the total number of calories if someone ate **two** of these muffins? [1 mark]
   (d) Suggest a topping for this muffin which would provide an increase in the protein content, and explain how this would help provide more protein. [4 marks]

Nutrition Facts	
Serving Size 1 muffin	
Serving Per Container 2	
Amount Per Serving	
Calories 250	Calories from Fat 110
	% Daily Value*
Total Fat 12 g	18%
Saturated Fat 3 g	15%
Cholesterol 30 mg	10%
Sodium 470 mg	20%
Total Carbohydrate 31 g	10%
Dietary Fiber 3 g	12%
Sugars 5 g	
Protein 5 g	

Figure 15.1 Food label for a pack of two English muffins

Structured questions

These questions usually give a piece of information, and ask you questions about it.

An example could be as follows:

2 Look at the recipe below and then answer the questions.

Pizza Margherita

Base

300 g strong bread flour

1 sachet quick yeast

1 tsp olive oil

pinch of salt

200 ml warm water.

Tomato sauce

100 ml tomato passata

1 clove garlic

handful of basil leaves

Topping

125 g ball mozzarella cheese, sliced

25 g parmesan cheese, grated

handful of basil leaves

20 cherry tomatoes cut in half

Method

1 Mix the base ingredients together to form a soft dough.
2 Knead for 10 minutes and leave in a warm place to rise for 20 minutes.

3 Make the sauce by crushing the garlic into the passata, then chopping the basil leaves finely and mix all three ingredients together. Leave in the bowl until ready.
4 Pre heat the oven to 220 degrees celsius.
5 Roll out the dough into two small thin rounds, or one large thin round, and place on lightly greased pizza trays or baking sheets.
6 Spread the sauce over the dough.
7 Scatter the halved tomatoes, mozzarella cheese and parmesan over the sauce.
8 Bake in the oven for ten minutes.
9 Tear the basil leaves and sprinkle on top of the pizza.
10 Serve with a salad.

(a) Identify which ingredients are **not** suitable for someone who is lactose intolerant, and explain why they are not suitable. [3 marks]
(b) Suggest **one** way to increase the fibre (NSP) content of this recipe. [1 mark]
(c) Explain why it is important to use strong bread flour for the dough. [2 marks]
(d) Explain why the dough is left to rise for 20 minutes. [3 marks]
(e) Suggest **two** extra vegetables that could be added to increase the vitamin content of this pizza. Explain which vitamins are contained in your selection of vegetables, and why these vitamins are important. [4 marks]

To answer this question:

- **Ensure** your answers are clear and factual. Part (a) would be answered with a factual reason why these ingredients are not suitable, and not just an answer like 'because they are not good for them'.
- **Refer** back to the recipe or information when you answer the questions.
- **Always** have a go at the questions even if you are not sure.

Some sample answers to this question are shown below, along with some commentary on how you might be awarded marks.

One mark is awarded for this point.

One mark is awarded for this explanation.

One mark is awarded for this.

(a) Answer: *The mozzarella cheese is not suitable for someone who is lactose intolerant as it contains milk sugar lactose, which will cause the sufferer to have stomach pains, bloating and excess wind, making them very uncomfortable. The parmesan cheese is not suitable for the same reasons.*

Commentary: In this question you are being asked to apply your knowledge and understanding of how specific dietary needs (in this case lactose intolerance) influence food choice. The question asks you to **identify** ingredients not suitable for someone with lactose intolerance, as well as to **explain** why they are not suitable. There are three marks available for this question – you will be awarded one mark for correctly identifying each unsuitable ingredient (mozzarella cheese and parmesan cheese), and a further mark for explaining why these ingredients are unsuitable. You need to give a **factual** reason – an answer like 'because they are not good for them' would not be awarded any marks.

(b) Answer: *The strong bread flour should be replaced with strong wholemeal flour to increase the fibre content.*

Commentary: In this question you are being asked to demonstrate you knowledge and understanding of how to adapt recipes to meet a specific nutritional need (in this case an increase in fibre content). There is just one mark available for this question and therefore you only need to make one point.

One mark is awarded for correctly identifying that strong bread flour contains gluten, which forms during the kneading process.

One mark is awarded for an explanation of the function of gluten in bread making.

(c) Answer: *Strong bread flour contains gluten. Gluten is the protein that is stretched during the kneading process of making bread, forming long strands of protein. As the bread rises the gluten stretches, trapping the air in the risen bread. When the bread is cooled, the gluten hardens as it denatures in the heat, and forms the structure of the bread.*

Commentary: This question is asking you to apply your knowledge and understanding of the working characteristics and functional and chemical properties of food (in this case flour). The question asks you to **explain** why strong bread flour should be used. There are two marks available for this question; you will be awarded full marks for a full explanation of why the flour should be strong.

(d) Answer: *As the bread dough is left in a warm place to rise, the yeast begins to grow and multiply, producing carbon dioxide gas. The gas is trapped in the dough by the stretched gluten, forming pockets of air which causes the dough to rise. This produces a softer, more pliable dough which will make a better structured base for the pizza.*

Two marks are awarded for the explanation of the fermentation process (what happens to the yeast when the dough is left to rise).

An additional mark is given for the explanation of why the dough is left to rise.

Commentary: This question is asking you to demonstrate and apply your knowledge and understanding of the working characteristics and functional and chemical properties of food (in this case dough used for bread making). The question asks you to **explain** why the dough needs to be left to rise. There are three marks available for this question, and to gain full marks you will need to give a complete explanation of the process. The student answer above would be awarded full marks.

One mark is awarded for naming carrots as a source of vitamins.

(e) Answer: *Carrots could be added, which contain vitamin A. Vitamin A is important for keeping your eyes in good health. Cherry tomatoes could be added for vitamin C. Vitamin C is important for healthy skin and gums, and for helping the absorption of non-haem sources.*

Another mark is awarded for naming vitamin A as present in carrots and for explaining its function.

One mark is awarded for naming cherry tomatoes as a source of vitamins.

Commentary: This question is asking you to demonstrate and apply your knowledge and understanding of the sources and functions of vitamins in the diet. It asks you to **suggest** two vegetables that would increase the vitamin content of the pizza, to **explain** which vitamins are contained in the vegetables and why they are important. You are awarded two marks for naming two extra vegetables and two additional marks for providing a full explanation of the vitamins included. The answer above would be awarded full marks. Other vegetables that could have been named include peppers (vitamins A and C), peas (vitamin c), and spinach (vitamin C or vitamin B2), but you would gain marks for naming any two suitable vegetables and a further two marks for full explanations of the vitamins they contain and their functions.

Another mark is awarded for naming vitamin C as present in cherry tomatoes and for explaining its function.

Free response questions

These questions ask you to write in more detail about a particular topic, but allow you to use your own knowledge to provide facts, examples and opinions to write in a way that you decide. These types of questions may be marked in bands, and you may have to give more answers than the number of marks shown to get maximum marks.

These questions will start with one of the following words from Table 15.1:

- **Describe**
- **Explain**
- **Analyse**
- **Evaluate**
- **Discuss.**

You will need to present the facts, opinions and examples in a structured way.

Always make a plan at the top of the paper so you can identify all the facts you want to include. That way you will not forget anything.

An example of this type of question is as follows:

3 **Protein** is a macronutrient.
  (a) Discuss why protein is important in the diet, detailing the functions of protein. [6 marks]

  (b) Explain why it is important for **a vegan** to eat a combination of protein to maintain a healthy body. [4 marks]

To answer this question:

- **Make a plan** to include everything you want to put in your answer. This could be a list or a spider diagram.
- **Keep referring** back to the question to make sure you are focussed on it.
- **Check** you are using scientific or relevant terms to increase the marks you can achieve.
- **Always** have a go at these questions even if you are not sure.

Some sample answers to this question are shown below, along with some commentary on how you might be awarded marks.

Protein provides chemicals for growth

Protein helps to repair the body

Protein helps to maintain the body, producing enzymes

Protein produces hormones to regulate body functions

Protein is a secondary source of energy.

(a) Answer: *Protein provides all the chemicals to make the body grow, particularly in children and pregnant women. It provides all the chemicals to help the body repair any damage after illness, accidents and surgery. It maintains the body to keep it working well, producing enzymes for digestion, muscle activity and nerve function. It provides chemicals to produce hormones, which regulate some body functions. It provides a secondary source of energy for the body.*

Commentary: This question is asking you to demonstrate and apply your knowledge and understanding of protein. It asks you to **discuss** why it is important in the diet and to **detail** the functions of protein. The table shows an example of how you might be awarded marks for this question.

Band	AO2 Maximum 6 marks
3	**Award 5–6 marks** An excellent response showing clear knowledge of the functions of protein, with reference to why protein is needed in the diet. The correct technical words are used confidently.
2	**Award 3–4 marks** A good response that includes reference to some of the functions of proteins. Reference is made to why protein is needed in the diet. Some correct terminology is used.
1	**Award 1–2 marks** A limited response with fewer than three suggestions for functions of protein. Little reference as to why protein is needed in the diet. There is limited use of correct terminology
0	**Award 0 marks** Response contains no information worth a mark.

(b) **Answer:** *All animal foods and soya beans contain the nine essential amino acids needed by the body. They are called High Biological Value (HBV) proteins. Foods containing Low Biological Value (LBV) proteins do not contain all the essential amino acids that are needed by the body. This means that if the body is to get all the essential amino acids from LBV foods (a vegan diet without animal foods does not include HBV sources of protein), a combination of these foods must be eaten. Foods that contain Low Biological Value (LBV) proteins are cereals (rice, oats, quinoa, wheat, millet) peas, beans (except soya beans), lentils, nuts and seeds. When these LBV foods are combined in meals they are called complementary proteins. Examples of meals that can be made from LBV foods to complement each other are beans on toast, hummus and pitta bread and rice and lentil dahl.*

Two marks are awarded for this explanation of LBV proteins and the vegan diet.

An additional two marks are awarded for the explanation of complementary proteins and examples of meals that could be included in the vegan diet.

Commentary: This question is asking you to demonstrate and apply your knowledge and understanding of proteins and how to plan a balanced diet for specific lifestyle needs (in this case a vegetarian). The question asks you to **explain** why a vegan needs to eat a combination of proteins for a healthy body. There are four marks available for this question and you will be awarded maximum marks for providing a full explanation of LBV proteins, including suggestions of meals containing complementary proteins. A simple explanation of complementary proteins would only allow you to gain two marks.

The table shows an example of how you might be awarded marks for this question.

Band	AO2 Maximum 4 marks
2	**Award 3–4 marks** A good response that shows clear knowledge and understanding of the importance of combining proteins in a vegan diet. Several suggestions for food combinations are given. The candidate has made good use of technical vocabulary.
1	**Award 1–2 marks** A limited response that shows some knowledge and understanding of complementary proteins, but little explanation as to why it is important to maintain a healthy body in a vegan diet. The candidate has attempted to use technical terminology.
0	**Award 0 marks** Response not creditworthy or not attempted.

Questions carrying the highest number of marks will ask you to analyse and evaluate different aspects of nutrition, food, cooking and preparation.

An example of this type of question is as follows:

4 The UK government has been trying to improve the health of UK citizens by recommending ways to improve the eating habits of people. Select one of the initiatives that the government has introduced. Explain what the initiative is and how it would help to improve a diet. Assess the impact of the suggestions made by the government. [8 marks]

A sample answer to this question is shown below, along with some commentary on how you might be awarded marks.

The student has identified a government initiative to improve healthy eating habits. 1 mark is awarded for this.

(a) *One of the ways that the government has tried to show people how to eat healthily is to have a diagram of a food plate with the amount of food types from food groups that people should be eating. This is called The Eatwell Guide. The Eatwell Guide is a visual representation of how different foods contribute towards a healthy balanced diet and replaces the Eatwell Plate.*

The foods are put into segments on the plate, and each segment contains food groups that are consistent with government recommendations for a diet that would provide all the nutrients required for a healthy adult or child over the age of 5.

The Eatwell guide is, based on showing food groups, making healthy eating easy to understand by showing people how much of each food group should be on a dinner plate.

People are shown that they should choose a variety of foods from the food groups of

Fruit and vegetables: These provide many of the essential vitamins and minerals needed, and also give fibre which is needed to keep the digestive system working correctly and prevent constipation. Around the outside of this section the information about eating at least five portions of a variety of fruit and vegetables every day is included.

Bread pasta and rice, and other starchy carbohydrates: These provide energy. They are a slow-release energy source, so fill you up and take a long time to digest. They can also provide fibre. The information about choosing wholegrain or higher fibre versions with less added fat, salt and sugar is included around the outside of this section.

Dairy and alternatives: These are a good source of protein and calcium. Protein is needed for growth and repair, and calcium is needed for strong bones and teeth. The information about choosing lower fat and sugar options is included around the outside of this section.

Beans, pulses, fish, eggs, meat and other proteins is a smaller pink section of the Eatwell Guide: Protein is essential for growth, repair and enzymes and hormones in the body. Animal foods and soya contain all the essential amino acids and are HBV proteins. The information about eating more beans and pulses, 2 portions of sustainably sourced fish per week, one of which is oily and eating less red and processed meat is included around the outside of this section.

There is a very tiny section labelled oil and spreads, which recommends that unsaturated oils are used in small amounts. This is because some fat is important in a healthy diet, but reflects the fact that oils and spreads are high fat and contain a lot of calories, so should be only consumed in small amounts.

There is information about the traffic light labelling on product and the need to keep hydrated and limit fruit juice as it is high in sugar content.

There is also a set of pictorial images showing foods to eat less often and in small amounts, which include cakes, chocolate, crisps, ice cream, biscuits and sweetened sauces.

> The student has then explained what the initiative is and how it would help improve a diet. 5 marks are awarded for this.

> *This new Eatwell Guide replaces the former Eatwell Plate. I think it shows a much better representation of what we should be eating now. I think that the Eatwell Plate had become out of date, as it included a picture of a can of coke and other foods that were not very healthy. I do not think the Eatwell Plate had the impact it should, as we still have many people who are overweight and obese, and the numbers are rising. Obesity also increases the risk of diabetes, so this is also on the rise in the UK.*
>
> *There are so many families who do not seem to take notice of these recommendations. It has been around for a number of years, so it was time for the Government to think of something new to encourage to people to think about their weight, and suggest changes that people can make to the things that they eat. However, people are not going to change their habits easily. The availability of fast food, ready meals that contain sugar and fats, people not being bothered to cook from fresh food and ingredients all contribute to the problem.*

The student has then assessed the impact of the suggestions made by the government, giving their opinions. 2 marks would be awarded for this.

Commentary:

In this question you are being asked to **analyse** and **evaluate** the impact the government's initiatives have had on people's health. You have been asked to **identify** one of the initiatives, **explain** what it is and how it would improve a diet, and give your opinions on the impact it has had on the health of people, using your knowledge.

There are **8** marks for this section of the question. You would be expected to fully explain the initiative, as above, including at least **four points** with a full explanation. Then you can identify what the initiative is trying to address and how the ideas suggested would improve a diet.

Your opinion can include any relevant information that backs up your ideas. The answer above includes information about the facts that obesity is still on the increase, and diabetes rates are rising.

Other possible answers could be:

I. The Eight Government steps as listed below, with explanations of each one.
    1   Base your meals on starchy foods.
    2   Eat a minimum of five portions of fruit and vegetables every day.
    3   Eat at least two portions of fish a week, one of which should be oily fish.
    4   Cut down on saturated fat and sugar.
    5   Eat less salt, no more than 6 grams a day for adults.

6 Drink plenty of water.

7 Do not skip breakfast.

8 Get active and try to be a healthy weight

These points would then be expanded and explained, followed by your opinion on whether this initiative has been successful, backed up with relevant information from your own knowledge of the health of UK citizens.

2 You could also discuss the **Five a day campaign** to encourage people to eat a minimum of five pieces of fruit and vegetables per day. You would need to explain why fruit and vegetables are so important, listing vitamins and minerals found in fruit and vegetables, what their functions are and what happens if we don't eat enough of these. You would have to advise on the size of each portion, explain why each vegetable is only counted once, e.g. why you cannot drink five glasses of orange juice and that is the five a day recommended portions.

The table shows an example of how you might be awarded marks for this question.

Band	AO4 Maximum 8 marks
3	**Award 6–8 marks** Excellent analysis and evaluation of the impact on health of the chosen Government initiative. Nutritional knowledge is excellent and within the response 3–4 points have been considered in depth and accurately interpreted in order to make judgements. Examples have been included. There is excellent use of correct terminology.
2	**Award 3–5 marks** A good attempt to analyse and evaluate the impact on health of the chosen Government initiative. Nutritional knowledge is good and 2–3 reasons have been addressed within the response. The selected facts have been adequately analysed and interpreted in order to make judgements. The candidate has used examples within the response. There is good use of correct terminology.
1	**Award 1–2 marks** A limited response that discusses some of the benefits of the chosen Government initiative. The response has made basic reference to 1–2 reasons. There is some attempt to analyse and interpret reasons and some basic judgements have been made. There is limited use of technical terminology.
0	**Award 0 marks** Response not creditworthy or not attempted.

GLOSSARY

Additives: substances added to food to preserve flavour, enhance taste and appearance, or extend shelf life.

Aeration: the adding of air during the combining of different ingredients.

Allergy: when someone suffers an unpleasant, sometimes life-threatening, reaction to a certain food or group of foods.

Amino acids: small units that join together to make large molecules of proteins.

Anaemia: the condition that occurs when you are not making enough red blood cells to carry oxygen around your body, so you will become tired.

Anaerobic: do not need oxygen.

Angina: a condition when the sufferer will experience chest pain and shortness of breath due to the arteries to the heart being partially blocked.

Bacteria: extremely small single-celled organisms that can only be seen under a microscope. Bacteria are found everywhere including in air, food, water and on animals and humans. They contaminate food and make it unfit to eat. Bacteria are able to reproduce very quickly by dividing into two about every 20 minutes if the conditions are correct.

Basal Metabolic Rate (BMR): the amount of energy needed to ensure the body functions are carried out before you do any physical activity.

Best-before date: the date by which it is recommended that the product is used, when it is of its best quality.

Biodegradable: a product or material that can dissolve easily in the environment without destroying nature.

Blanch: vegetables or fruit are put into boiling water for two minutes and then plunged into iced water. This helps prevent enzymic browning and partly cooks the vegetables or fruit.

Body Mass Index (BMI): the calculation used to find out whether you are a healthy weight. A healthy BMI is between 18.5 and 24.9.

Bridge hold: the safe way to hold small fruit and vegetables while cutting into pieces, making your hand into a bridge across the vegetable of fruit and cutting between your fingers and thumb.

Caramelisation: when sugars melt at high heat to change colour to a shade of brown, and release sweetness.

Carbon footprint: the amount of carbon emissions that have been produced during the growing, processing and distribution (transporting) of a food product.

Cardiovascular disease: any disease of the heart or blood vessels.

Cellulose: insoluble substance which makes up most of plants, like vegetables' cell walls.

Cholesterol: the fatty substance made in the liver which is carried in the blood. It exists in bad forms (LDL) and good forms (HDL).

Claw grip: the safe way to hold fruit and vegetables when slicing them into small pieces, by making a claw of your fingers to guide the knife.

Coagulation: this is when protein denatures and forms a solid structure.

Coeliac disease: an intolerance to the protein called gluten, which causes the inflammation of the intestine walls and damage to them, making nutrient absorption more difficult for the body.

Colloidal foam: when two ingredients are mixed together to form a foam, e.g. eggs and sugar.

Complementary actions: how different nutrients work with each other in the diet.

Complementary proteins: LBV proteins that are eaten in one meal together to provide the essential amino acids.

Concentrated: a richer, more intense food source due to removal of water.

Conduction: this is where the heat is directly touching a food, and the heat gives energy to the molecules in the food, which start to vibrate and pass energy and heat through the food to cook it.

Connective tissue: a protein called collagen found in meat.

Conserving nutritive value: how to stop nutrients being lost from foods during the cooking process.

Contamination: the presence of harmful micro-organisms or chemicals in food.

Convection: this happens in liquids and gases. The heat gives molecules in the liquid or gas energy, they rise and colder molecules take their place. This creates convection currents.

Coronary heart disease: when the arteries supplying the heart with oxygenated blood become partially blocked with fatty deposits from cholesterol, so the blood cannot flow freely.

Cross-contamination: the transferring of bacteria from one source to another.

Cuisine: a style characteristic of a particular country or region with which specific ingredients, cooking methods, equipment and presentation and serving styles are associated.

Cultivated: raised or grown on a farm or grown under controlled conditions.

Curing: when meat is treated with chemicals like sodium nitrite to preserve it for longer periods of time.

Date mark: date labels added to products. Use-by date and best before date are both date marks.

Denature: the unravelling of the bonds that hold amino acids together in proteins, and the creation of a different structure of amino acids.

Dental caries: the rotting of teeth caused by a diet with too much sugar and insufficient cleaning of the teeth.

Dextrinisation: this is when dry heat is applied to a starchy product, and the molecules on the surface break down and change colour to brown.

Dietary Reference Values (DRVs): the amount of a nutrient a person needs.

Digestible: some foods are more quickly broken down by the digestive system than others. These foods are referred to as more digestible foods: food that can be broken down during the digestive process (by the action of enzymes) into the macronutrients or micronutrients that can be used by the body. These are absorbed by the body through the wall of the intestine.

Disaccharides: double molecules of glucose joined together. They take longer to digest. They include sucrose, lactose and maltose.

Dormant: not active, asleep but not killed.

Dovetailing: where your plan shows you are in charge of more than one thing at a time, and you are aware that one thing is cooking while you are making something else. This is multi-tasking.

Dredging: towing metal cages across shellfish beds to catch shellfish.

Durum wheat: a high biological value protein containing wheat.

Emulsification: the property of fat that allows it to be mixed with a liquid after the addition of an emulsifier, such as egg yolk.

Endosperm: a tissue produced inside seeds of flowering plants around the time of fertilisation. It surrounds the embryo and provides nutrition in the form of starch, though it can also contain oils and protein.

Energy balance: where we take in as much energy in the food we eat as we are using up in our daily activity.

Enzymes: protein molecules that control chemical reactions in food.

Enzymic browning: The release of enzymes from cut cells in fruit or vegetables, which react with oxygen and form brown discolourations on the product.

Essential amino acids: the nine amino acids that cannot be made by our bodies, so we must eat the proteins that contain them.

Essential fatty acids: small units of fat needed to keep our bodies functioning properly.

Estimated Average Requirement (EAR): the amount of calories needed per day for your body to work properly and to maintain your body weight. It is calculated by BMR x PAL.

Extraction rate: how much of the whole grain has been used in products.

Extrinsic sugars: sugars that we add to recipes and dishes we make and drinks.

Extruded: a soft mixture being squeezed under pressure through a shaped die into strips and cut into even pieces.

Factory farms: these farms maximise the number of animals that can be reared. Factory farming focuses on profit and efficiency rather than the welfare of the animals.

Farmed fish: fish raised in tanks, sea enclosures or cages.

Fat-soluble vitamins: vitamins that dissolve in fat. This includes vitamins A, D, E and K.

Fermented: when milk has been turned into a different food like cheese or yoghurt using harmless bacteria which enable the sugar to convert into lactic acid.

First in first out: a rule used when storing food, so that older items are used first.

Flesh: area under the potato skin.

Foam formation or aeration: where whisking or beating an egg will result in the protein structure changing to trap small bubbles of air.

Food miles: the distance the food has travelled from field or production to the consumer's plate.

Food poisoning: an illness that is caused by consuming food or water that has been contaminated by specific pathogenic bacteria or their toxins, or by other harmful micro-organisms.

Food security: when all people at all times have access to sufficient, safe and nutritious food, to maintain a healthy and active life.

Food spoilage: when food deteriorates so that its quality is reduced and/or it is inedible.

Fortified: foods which have extra nutrients added, either to replace those lost in manufacture/food processing, or to compensate for a missing nutrient present in a traditional alternative.

Gelation: solidifying a mixture by chilling or freezing.

Global warming: the gradual heating of the Earth's surface, oceans and atmosphere.

Gluten: the protein found in wheat, barley, oats and rye products.

Halal: foods that Muslims are allowed to eat.

Haram: foods that Muslims are not allowed to eat.

Harvested: gathering or picking of crops.

High Biological Value (HBV) proteins: foods that contain all the essential amino acids.

High-risk food: foods which have ideal conditions for the growth of bacteria. They are often foods high in protein and moisture.

High-yield: grow in abundance; grown to produce large results.

Hydrogenation: the process of turning oils into solid fats.

Hydrophobic: a product, such as fat, that will not mix with water.

Imported: to bring foods into a country from abroad for sale.

Infusion: the flavours extracted from any ingredient by soaking or steeping in water or oil.

Insoluble fibre: absorbs water and helps prevents constipation.

Intrinsic sugars: sugars contained within plant cells.

Invisible fat: fat that cannot be easily seen and is impossible to separate from food. It is found in products we eat, such as biscuits, ice cream and ready meals.

Jus: the liquid obtained from the cooking of an ingredient, often meat.

Kosher: foods that Jewish people can eat.

Lacto vegetarian: a person who does not eat meat, fish or eggs, or products containing these, but who will eat cheese and dairy products.

Lacto-ovo vegetarian: a person who has chosen not to eat any meat or fish, or products containing these, but will eat eggs, cheese and dairy products.

Lactose intolerance: when a person is unable to digest the sugar found in milk, which is called lactose. They cannot eat any dairy products which contain lactose.

Lactose: a sugar from carbohydrates found in milk.

Laminated: many different layers of materials bonded together.

Leavened: risen. Yeast is usually the raising agent.

Lingual papilla: the raised hair-like structures on your tongue that contain the taste buds.

Lipids: a general term given to fats.

Low Biological Value (LBV) proteins: foods that contain some of the essential amino acids.

Macronutrients: nutrients needed by the body in large amounts. They include proteins, fats and carbohydrates.

Maillard reaction: chemical reaction between proteins and carbohydrates, which changes the flavour of the food.

Marinate: to soak meat, fish or alternative proteins in a weak acid solution, such as lemon juice or vinegar, to denature the protein and make it more tender.

Micronutrients: nutrients needed by the body in small amounts. They include vitamins, minerals and trace elements.

Minerals: help to make strong bones and teeth, make sure we have sufficient red blood cells to transport oxygen around the body, control the amount of water in our body and make the nerves and muscles work correctly. They include calcium, iron, magnesium, phosphorus, potassium and sodium.

Monosaccharides: simple sugars made of small molecules that are easily digested. They include glucose, fructose and galactose.

Monounsaturated fats: fats containing a pair of carbon atoms with only one hydrogen atom attached. They are soft at room temperature, but will harden when put in the fridge. They are considered to be healthier than other fats.

Moulds: a type of fungi. Moulds reproduce by producing spores, which will travel in the air and land on food.

Non Starch Polysaccharides (NSP): also called fibre. The non-digestible part of plant cell walls. It provides bulk in the diet and helps to move the waste food through the system.

Olfactory cortex: the part of the brain that deals with smells and aromas.

Olfactory system: the area of your nose that detects smells and aromas.

Organic farms: these farms put the welfare of the animal first and there are certain standards that have to be met.

Osteoporosis: a condition that develops when the creation of new bone does not keep up with the removal of old bone, and the bones become weak and are at risk of fracturing and breaking easily.

Oxidation: the reaction of oxygen with the cut surface of a fruit or vegetable.

Palatability: the way the food tastes and its appeal to us.

Pasteurisation: a method of heat-treating milk using a high heat and only seconds in time.

Peak bone mass: the amount of bony tissue present when your skeleton has stopped growing and your bones are at their maximum strength and density.

Perish: to spoil or to go bad.

Pernicious anaemia: caused by low red blood cell production, which happens when the body is unable to absorb vitamin B12. Vitamin B12 is needed, with iron, to produce red blood cells.

Physical Activity Level (PAL): the amount of extra activity you do per day, including sitting, standing, running and any extra exercise you carry out during the day.

Pith: the watery core.

Plant-based non-haem iron: in animal foods, iron is attached to proteins called haem proteins, and it is easier to absorb. In plant foods, the iron is not attached to haem proteins, and it is harder for the body to absorb the iron.

Plasticity: the property of fat which allows different fats to melt at different temperatures.

Polysaccharides: complex carbohydrates made of long chains of sugar molecules that take a long time to digest. They include starch, fibre (NSP), pectin, dextrose and glycogen.

Polytunnel: large plastic tunnel/ frame to cover foods.

Polyunsaturated fats: have two or more pairs of carbon atoms which are capable of taking up more hydrogen atoms. They are soft and oily at room temperature and will not harden in the fridge.

Porous: contains tiny holes which allows moisture and gases to pass through.

Primary food: a food that is not edible in its original state; it has to be changed or have some form of preparation before it can be eaten. Examples include raw potato, wheat and maize.

Primary processing: the process of changing or transforming primary foods into a suitable state to either be eaten or used in the production of other products.

Processed: changes made to a raw material to make it more edible.

Protein denaturation: where protein strands unravel and change structure to form a different shape or colour.

Purse-seining: drawing a huge net around a school of fish (for example, tuna, herring, mackerel) to catch them.

Radiation: happens in grilling and microwaving, when infra-red waves or microwaves pass through the air

and heat the surface of the food, giving the molecules energy.

Rancid: to have an unpleasant smell or taste sour.

Reared: the conditions an animal is raised in when young.

Recommended Daily Intake (RDI): the amount of each nutrient calculated to be sufficient to meet the requirements of the majority of the population of the UK.

Reduction: when the water is allowed to evaporate from a sauce or liquid, concentrating the flavours.

Ripened cheese: a cheese that has an external mould or a bloomy rind on the outside.

Satiety: feeling full after eating.

Saturated fats: fats with two hydrogen atoms for each carbon atom that are mainly solid at room temperature and mainly animal fats.

Seasonal or Seasonality: the times of year a particular food product is available because of the time it ripens. The food is also usually at its cheapest price at this time. An example is British asparagus, which is only available in May and June, because that is when it is ready to harvest.

Secondary processing: when a primary food is changed or converted into an ingredient which can then be used to make a food product.

Sensory perception: the way we recognise flavour in food, combining taste, smell, feel and sight.

Sensory properties: the characteristics of a food that are linked to the smell, sight, taste and feel.

Shelf life: the length of time a product may be stored without becoming unsuitable to use or eat.

Shortening: the ability of a fat to produce a characteristic crumbly texture to baked products (when flour is coated with fat to prevent gluten formation).

Skin: the outside layer or covering.

Soluble fibre: slows down the digestive process and can help lower blood cholesterol levels.

Somatosensory cortex: the part of the brain that deals with touch.

Sous vide: a cooking method in which food is cooked slowly at low temperatures in a vacuum-sealed pouch.

Starch degradation: the breaking up of starch molecules during cooking, which releases sugars and sweetness, and breaks down bonds to make the starchy food softer.

Starch gelatinisation: where starch molecules swell up during boiling and absorb water to thicken a sauce. They eventually burst, creating a gel in the sauce.

Starter culture: combined single strains of bacteria.

Sugar beet: a root crop similar to a parsnip grown in a field in countries with warm and cold seasons.

Sugar cane: a tall grass grown in hot countries.

Tampering: to interfere with food to cause damage.

Taste receptors: these are located at the top of the taste buds and send information via the nervous system to the brain so that flavours are identified.

Tenderise: to hit meat with a mallet to break down the bonds between the amino acids and help make the meat more tender.

Toxic: containing a harmful substance.

Toxins: poisons that can cause illness.

Trace elements: responsible for strengthening the tooth enamel, making hormones and controlling bodily functions, and act in other muscle and nerve functions. They include fluoride and iodine.

Trans-fatty acids: manmade molecules created when manufacturers add hydrogen to vegetable oils (hydrogenation).

Trawling: using a net or nets to catch fish.

Tuber: A thickened underground part of a stem or root from which new plants can grow.

Type 2 diabetes: the condition that develops when your body stops producing enough insulin to deal with the glucose levels in the blood.

UHT: ultra heat treatment – a method of heat-treating milk to kill bacteria and extend the shelf life.

Umami: one of the five elements of taste that has a savoury or meaty flavour.

Unleavened: flat, not risen.

Unripened cheese: a cheese that has a soft and smooth texture and does not have a rind.

Unsaturated fats: fats which are usually liquid or soft at room temperature.

Use-by date: the product is safe to consume up to and including this date.

Vegan: a person who will not eat any animal product, or foods containing any animal product.

Visible fats: fats that can be seen, such as the fat on meat, butter or oils that we use for frying or salad dressing.

Water-soluble vitamins: vitamins that dissolve in water. This includes vitamin B group and vitamin C.

Weevils: tiny black bugs that can live and breed in flour.

Wholegrain: the whole seed in its natural state, none of the layers have been removed.

Yeasts: one-celled organisms which come from the group of organisms called fungi. Yeasts reproduce by budding; this means they multiply and the one cell divides into two.

INDEX

NOTES: page numbers in **bold** indicate Keywords